My Book

This book belongs to

Name: _____

Grade 8
Work book

Index

Grade 8 Work book

Index

Copy right © 2019 MATH-KNOTS LLC

All rights reserved, no part of this publication may be reproduced, stored in any system or transmitted in any form, or by any means, electronic, mechanical, photocopying, recording, or otherwise without the written permission of MATH-KNOTS LLC.

Cover Design by :
Gowri Vemuri

First Edition :
March , 2023

Author :
Gowri Vemuri

Edited by :
Raksha Pothapragada
Ritvik Pothapragada

Questions: mathknots.help@gmail.com

NOTE : CCSSO or NCTM is neither affiliated nor sponsors or endorses this product.

Grade 8
Work book

Index

Dedication

This book is dedicated to:
My Mom, who is my best critic, guide and supporter.
To what I am today, and what I am going to become tomorrow,
is all because of your blessings, unconditional affection and support.

This book is dedicated to the
strongest women of my life ,
my dearest mom
and
to all those moms in this universe.

G.V.

Grade 8 Work book

Index

Grade 8 Work book

Index

Math-Knots Grade level enrichment series covers all pre-K to Grade 10 common core math work books.

The Grade 6 math work book is aligned to common core curriculum and more, challenge level questions are included. Content is divided across weeks aligned to school calendar year.
Six assessments are provided based on the topics covered in the previous weeks. The assessments help them to identify learning gaps students can redo the previous weeks content to bridge their learning gaps as needed. End of the year assessment is provided online at the below URL.
www.a4ace.com

For more practice ,you can also subscribe at www.a4ace.com .
All practice sets, and assessments can be taken any number of times within the one-year subscription period.
All our content is written by industry experts with over 30 years of experience.
A4ace.com is part of Math-Knots LLC
Math-Knots is your one stop enrichment place. ~~Learn to think with us.

Note: Video explanations of content in the book is coming up.

Instructions to register on www.a4ace.com
1. Register as a parent and choose a category while registering.
2. Register the student and choose a category while registering.
3. You can add more categories from the dash board as needed.
4. After registering an automatic email will be sent for verification.
5. Verify the account by clicking on the link sent to your email.
6. Login with your credentials and navigate to the dash board.
7. Click on the free test and follow through the instructions.
8. Free test will navigate to a payment page and it will say $0.
   You just have to click Paypal button and you can take the test.
9. Any issues please reach out to mathknots.help@gmail.com

Grade 8 Work book

Index

WEEK #	CHAPTER	TOPIC		PAGE #
Week 1, 2 & 3	Number System	Rational Vs Irrational Numbers	8.NS1 8.NS2	21 - 64
	Expressions & Exponents	Exponents	8.EE.1	
Week 4			8.EE.2	65 - 76
Week 5		Square and Cube roots	NS1.2	77 - 91
Week 6 & 7		Scientific Notations	8.EE.3 8.EE.4	92 - 115
Week 8	ASSESSMENT # 1			116 - 124
Week 9	Equations	Slopes and Missing coordinate	8.EE.5 8.EE.6 F.4	125 - 152
		Graphing Equations	F.5	
		Parallel & Perpendicular lines		153 - 169
Week 10		Two step Linear Equations		
Week 11		Two step Linear Equations	8.EE.7	170 - 184
Week 12		Distance between points Mid point		185 - 201
Week 13		Simultaneous Equations	8.EE.8	202 - 218
Week 14		Simultaneous Equations and Graphs	8.EE.8	219 - 259
Week 15	ASSESSMENT # 2			260 - 275
	Answer Keys			276 - 306

©All rights reserved-Math-Knots LLC., VA-USA

Real Number System

Real number system

Irrational numbers
$\sqrt{2}, \sqrt{3}, \sqrt{5}, \ldots$

Rational numbers

Integers

Whole numbers

Natural numbers
1, 2, ...

0, 1, 2, 3...

..., -3, -2, -1, 0, 1, 2, 3, ...

$-\infty \ldots -2, \frac{-3}{2}, -1, \frac{-1}{2}, \frac{-1}{3}, 0, \frac{1}{3}, \frac{1}{2}, 1, \frac{3}{2}, 2 \ldots \infty$

Grade 8

Laws of indices:

For all real numbers a and b and all rational numbers m and n, we have

(i) $a^m \times a^n = a^{m+n}$

Examples: (1) $2^3 \times 2^6 = 2^{3+6} = 2^9$

(2) $\left(\dfrac{5}{6}\right)^4 \times \left(\dfrac{5}{6}\right)^5 = \left(\dfrac{5}{6}\right)^{4+5} = \left(\dfrac{5}{6}\right)^9$

(3) $5^{2/3} \times 5^{4/3} = 5^{(2/3 + 4/3)} = 5^{6/3} = 5^2$

(4) $2^3 \times 2^4 \times 2^5 \times 2^8 = 2^{(3+4+5+8)} = 2^{20}$.

(5) $(\sqrt{7})^3 \times (\sqrt{7})^{\frac{5}{2}} = (\sqrt{7})^{3+\frac{5}{2}} = (\sqrt{7})^{\frac{11}{2}}$

(ii) $a^m \div a^n = a^{m-n},\ a \neq 0$

Examples: (a) $7^8 \div 7^3 = 7^{8-3} = 7^5$

(b) $\left(\dfrac{7}{3}\right)^9 \div \left(\dfrac{7}{3}\right)^5 = \left(\dfrac{7}{3}\right)^{9-5} = \left(\dfrac{7}{3}\right)^4$

(c) $9^{\frac{2}{3}} \div 9^{\frac{1}{6}} = 9^{\left(\frac{2}{3}-\frac{1}{6}\right)} = 9^{\left(\frac{4-1}{6}\right)} = 9^{\frac{3}{6}} = 9^{\frac{1}{2}}$

(d) $\left(\dfrac{5}{7}\right)^{\frac{8}{9}} \div \left(\dfrac{5}{7}\right)^{\frac{1}{3}} = \left(\dfrac{5}{7}\right)^{\left(\frac{8}{9}-\frac{1}{3}\right)} = \left(\dfrac{5}{7}\right)^{\frac{8-3}{9}} = \left(\dfrac{5}{7}\right)^{\frac{5}{9}}$

Note: $a^n \div a^n = 1$
or $a^{n-n} = a^0 = 1$
$\therefore a^0 = 1,\ a \neq 0$

(iii) $(a^m)^n = a^{m \times n}$

Examples: (a) $(5^2)^3 = 5^{2 \times 3} = 5^6$

(b) $\left[\left(\dfrac{2}{3}\right)^4\right]^5 = \left(\dfrac{2}{3}\right)^{4 \times 5} = \left(\dfrac{2}{3}\right)^{20}$

(c) $\left[\left(\dfrac{5}{7}\right)^{\frac{2}{3}}\right]^{\frac{9}{8}} = \left(\dfrac{5}{7}\right)^{\left(\frac{2}{3} \times \frac{9}{8}\right)} = \left(\dfrac{5}{7}\right)^{\frac{3}{4}}$

(iv) $\left(\dfrac{a}{b}\right)^n = \dfrac{a^n}{b^n}$

Example: $\left(\dfrac{4}{5}\right)^7 = \dfrac{4^7}{5^7}$

Note: Conversely we can write $\left(\dfrac{a^n}{b^n}\right) = \left(\dfrac{a}{b}\right)^n$

Example: $\dfrac{8}{27} = \dfrac{2^3}{3^3} = \left(\dfrac{2}{3}\right)^3$

(v) $(ab)^n = a^n \times b^n$

Examples: (a) $20^5 = (4 \times 5)^5 = 4^5 \times 5^5$
(b) $(42)^7 = (2 \times 3 \times 7)^7 = 2^7 \times 3^7 \times 7^7$

Note: Conversely we can write $a^n \times b^n = (ab)^n$

Examples: (a) $4^8 \times 5^8 = (4 \times 5)^8 = 20^8$
(b) $\left(\dfrac{2}{3}\right)^5 \times \left(\dfrac{9}{8}\right)^5 = \left(\dfrac{2}{3} \times \dfrac{9}{8}\right)^5 = \left(\dfrac{3}{4}\right)^5$

(vi) $a^{-n} = \dfrac{1}{a^n}$, $a \neq 0$

Example: $2^{-4} = \dfrac{1}{2^4}$, $5^{-1} = \dfrac{1}{5}$

Note: $a^{-1} = \dfrac{1}{a^1} = \dfrac{1}{a}$

(vii) $\left(\dfrac{a}{b}\right)^n = \left(\dfrac{b}{a}\right)^n$

Examples: (a) $\left(\dfrac{5}{9}\right)^3 = \left(\dfrac{9}{5}\right)^{-3}$

(b) $\left(\dfrac{1}{5}\right)^{-1} = \left(\dfrac{5}{1}\right)^1 = 5$

Note: $\left(\dfrac{1}{a}\right)^{-1} = \left(\dfrac{a}{1}\right)^1 = a$

(viii) If $a^m = a^n$, then m = n, where $a \neq 0$, $a \neq 1$

Examples: (a) If $5^p = 5^3 \Rightarrow p = 3$
(b) If $4^p = 256$
$4^p = 4^4 \Rightarrow p = 4$

(ix) For positive numbers a and b, if $a^n = b^n$, $n \neq 0$, then a = b (when n is odd)

Examples: (a) If $5^7 = p^7$, then clearly p = 5.
(b) If $(5)^{2n-1} = (3 \times p)^{2n-1}$, then clearly 5 = 3p or $p = \dfrac{5}{3}$

(x) If $p^m \times q^n \times r^s = p^a q^b r^c$, then m = a, n = b, s = c, where p, q, r are different primes.

Examples: (a) If $40500 = 2^a \times 5^b \times 3^c$, then find $a^a \times b^b \times c^c$

2	40,500
2	20,250
5	10,125
5	2,025
5	405
3	81
3	27
3	9
	3

$\therefore 40500 = 2^2 \times 5^3 \times 3^4 = 2^a \times 5^b \times 3^c$

$\therefore$ a = 2, b = 3, c = 4, [Using the above law].

$\therefore a^a \times b^b \times c^c = 2^2 \times 3^3 \times 4^4 = 27,648$

Example 8 : (a) $(20)^5 = (4 \times 5)^5 = 4^5 \times 5^5$
(b) $(42)^7 = (2 \times 3 \times 7)^7 = 2^7 \times 3^7 \times 7^7$

Note: Conversely we can write $a^n \times b^n = (ab)^n$

Grade 8

Vol 1 Notes

Example 9 : (a) $((5)^3)^2 = 5^{3 \times 2} = (5)^6 = 5 \times 5 \times 5 \times 5 \times 5 \times 5 = 15625$

(b) $(2)^3 \times (2)^5 = (2)^{3+5} = (2)^8 = 2 \times 2 \times 2 \times 2 \times 2 \times 2 \times 2 \times 2 = 256$

(c) $(7)^0 = 1$ | Any base value rise to the power zero is always equal to 1

(d) $(3)^{-4} = \dfrac{1}{(3)^4} = \dfrac{1}{3 \times 3 \times 3 \times 3} = \dfrac{1}{81}$

(d) $\dfrac{(8)^7}{(8)^5} = (8)^{7-5} = (8)^2 = 64$

(e) $\dfrac{(9)^4}{(9)^7} = (9)^{4-7} = (9)^{-3} = \dfrac{1}{(9)^3} = \dfrac{1}{9 \times 9 \times 9} = \dfrac{1}{729}$

(f) $(2)^4 = 2 \times 2 \times 2 \times 2 = 16$ (g) $(-2)^4 = -2 \times -2 \times -2 \times -2 = 16$

(h) $-(2)^4 = -(2 \times 2 \times 2 \times 2) = -16$ (i) $-(2)^3 = -(2 \times 2 \times 2) = -8$

(j) $(-2)^3 = -2 \times -2 \times -2 = -8$ (k) $-(-2)^3 = -(-2 \times -2 \times -2) = -(-8) = 8$

Tip 1 : When the exponent is an even number the simplified value is always positive, when the base has a positive or negative value.

Tip 2 : When the exponent is an odd number the simplified value is always positive, when the base has a positive value.

Tip 3 : When the exponent is an odd number the simplified value is always negative, when the base has a negative value.

Grade 8

Square Roots :

$\sqrt{a} - \sqrt{b} \neq \sqrt{(a-b)}$

$\sqrt{a} + \sqrt{b} \neq \sqrt{(a+b)}$

$\sqrt{ab} = \sqrt{a} \cdot \sqrt{b}$

$\sqrt{\dfrac{a}{b}} = \dfrac{\sqrt{a}}{\sqrt{b}}$

$\sqrt{a^2} = a$

$\sqrt[n]{a} = a^{\frac{1}{n}}$

Examples :

$\sqrt{8} - \sqrt{5} \neq \sqrt{(8-5)}$

$\sqrt{6} + \sqrt{4} \neq \sqrt{(6+4)}$

$\sqrt{4 \cdot 9} = \sqrt{4} \cdot \sqrt{9} = 2 \cdot 3 = 6$

$\sqrt{\dfrac{4}{9}} = \dfrac{\sqrt{4}}{\sqrt{9}} = \dfrac{2}{3}$

$\sqrt{5^2} = 5 = \sqrt{25}$

$\sqrt[n]{16} = 16^{\frac{1}{n}}$

$\sqrt[4]{16} = 16^{\frac{1}{4}} = (2^4)^{\frac{1}{4}} = 2$

Like Terms :

Two or more terms are said to be alike if they have the same variable and the same degree.
Coefficients of like terms are not necessarily be same.

An expression is in its simplest form when

1. All like terms are combined.
2. All parentheses are opened and simplified.

Like Terms can combined by adding or subtracting their coefficients (pay attention to the positive and negative signs of the coefficient and apply rules of adding integers)

Example 1 : -2x + 5x + 7 = 3x + 7

> Note : -2x and 5x are like terms and can be combined using rules of integers

Example 2 : -11y + 5 + 8y - 7 = -3y - 2

> Note : -11y and 8y are like terms and can be combined using rules of integers. 5 and -7 are like terms and can be combined using rules of integers

Example 3 : -12a - 5a + 8 - 3 = -17a + 5

> Note : -12a and -5a are like terms and can be combined using rules of integers. 8 and -3 are like terms and can be combined using rules of integers

Example 4 : -5b + 7 - 3b + 2a - a + 10 = -8b + a + 17

> Note : -5b and -3b are like terms and can be combined using rules of integers. 2a and -a are like terms and can be combined using rules of integers. 7 and 10 are like terms and can be combined using rules of integers.

Distributing with the negative sign :

Remember to apply the integer rules of positive and negative numbers while distributing.

Combining like terms on the opposite side of the equal sign :

When the like terms are on opposite sides, we have to combine like terms by using the inverse operation and by undoing the equation.

Example 5 : $-2x + 5 = -7x$

$$-2x + 5 = -7x$$
$$+7x \quad\quad +7x$$
$$\overline{-2x + 7x + 5 = -7x + 7x}$$
$$5x + 5 = 0$$

Solving equations using the distributive property :

The number in front of the parentheses needs to be multiplied with every term within the parentheses. After the distribution and opening up the parentheses, combine like terms and solve.

Distributing with the negative sign :

Remember to apply the integer rules of positive and negative numbers while distributing.

+ X + = +
- X - = +
- X + = -
+ X - = -

Example 6 :
(a) $2(5x + 7) = 2(5x) + 2(7) = 10x + 14$
(b) $-7(3a + 8) = (-7)(3a) + (-7)(8) = -21a + (-56) = -21a - 56$
(c) $3(-5b - 2) = (3)(-5b) - (3)(2) = -15b - 6$
(d) $6(-4a + 5) = (6)(-4a) + (6)(5) = -24a + 30$
(e) $4(2a - 8) = (4)(2a) - (4)(8) = 8a - 32$
(f) $-5(a - 7) = (-5)(a) - (-5)(7) = -5a - (-35) = -5a + 35$
(g) $-9(-2a + 10) = (-9)(-2a) + (-9)(10) = 18a + (-90) = 18a - 90$
(h) $-8(-5a - 6) = (-8)(-5a) - (-8)(6) = 40a - (-48) = 40a + 40a$
(i) $-(a + 7) = -a - 7$
(j) $-(x - 5) = (-1)(x) - (-1)(5) = -x - (-5) = -x + 5$
(k) $-(-a - b) = (-1)(-a) - (-1)(b) = a - (-b) = a + b$
(l) $-(-a + 2b) = (-1)(-a) + (-1)(2b) = a + (-2b) = a - 2b$

Example 7 : $2x + 3 = x + 7$
$2x + 3 = x + 7$
$\underline{-x - 3\ \ -x\ -3}$ ⟵ Inverse operation for addition is subtraction
$x + 0 = 0 + 4$
$x = 4$

Example 8 : $7x + 5 = -3x + 25$
$7x + 5 = -3x + 25$
$\underline{3x\ \ -5\ \ \ 3x\ \ \ -5}$ ⟵ Inverse operation for addition is subtraction and vice versa
$10x + 0 = 0 + 20$
$10x = 20$
$\dfrac{\cancel{10}x}{\cancel{10}} = \dfrac{\cancel{20}^{\,2}}{\cancel{10}}$ ⟵ Inverse operation for multiplication is division
$\boxed{x = 2}$

Example 9 : $\dfrac{2x}{5} + 5 = 15$

$\dfrac{2x}{5} + 5 = 15$
$\underline{-5\ \ \ -5}$ ⟵ Inverse operation for addition is subtraction and vice versa
$\dfrac{2x}{5} + 0 = 10$

$\dfrac{2x}{5} = 10$

$5 \cdot \dfrac{2x}{\cancel{5}} = 5 \cdot 10$ ⟵ Inverse operation for division is multiplication

$\dfrac{\cancel{2}x}{\cancel{2}} = \dfrac{\cancel{50}^{\,25}}{\cancel{2}}$ ⟵ Inverse operation for multiplication is division

$\boxed{x = 25}$

Grade 8

Vol 1
Week 1
Rational numbers

1) Which of the below is a rational number?

(A) $\sqrt{9}$ (B) $\sqrt{5}$
(C) $\sqrt{11}$ (D) $\sqrt{6}$

2) Which of the below is a rational number?

(A) $\sqrt{80}$ (B) $\sqrt{42}$
(C) $\sqrt{35}$ (D) $\sqrt{81}$

3) Which of the below is a rational number?

(A) $\sqrt{25}$ (B) $\sqrt{21}$
(C) $\sqrt{30}$ (D) $\sqrt{19}$

4) Which of the below is a rational number?

(A) $\sqrt{3}$ (B) $\sqrt{111}$
(C) $\sqrt{4}$ (D) $\sqrt{216}$

5) Which of the below is an irrational number?

(A) $\sqrt{7}$ (B) $\sqrt{16}$
(C) $\sqrt{49}$ (D) $\sqrt{900}$

6) Which of the below is a rational number?

(A) $\sqrt{32}$ (B) $\sqrt{49}$
(C) $\sqrt{41}$ (D) $\sqrt{99}$

7) Which of the below is a rational number?

(A) $\sqrt{21}$ (B) $\sqrt{63}$
(C) $\sqrt{88}$ (D) $\sqrt{49}$

8) Which of the below is an irrational number?

(A) $\sqrt{121}$ (B) $\sqrt{117}$
(C) $\sqrt{169}$ (D) $\sqrt{144}$

9) Which of the below is an irrational number?

(A) $\sqrt{81}$ (B) $\sqrt{64}$
(C) $\sqrt{400}$ (D) $\sqrt{5}$

10) Which of the below is an irrational number?

(A) $\sqrt{71}$ (B) $\sqrt{1089}$
(C) $\sqrt{961}$ (D) $\sqrt{196}$

©All rights reserved-Math-Knots LLC., VA-USA   www.math-knots.com | www.a4ace.com

Grade 8

11) Which of the below is an irrational number ?

(A) $\sqrt{225}$ (B) $\sqrt{25}$

(C) $\sqrt{23}$ (D) $\sqrt{36}$

12) Which of the below is an irrational number ?

(A) $\sqrt{256}$ (B) $\sqrt{39}$

(C) $\sqrt{16}$ (D) $\sqrt{289}$

13) Which of the below is an irrational number ?

(A) $\sqrt{441}$ (B) $\sqrt{686}$

(C) $\sqrt{484}$ (D) $\sqrt{84}$

14) Which of the below is an irrational number ?

(A) $\sqrt{36}$ (B) $\sqrt{121}$

(C) $\sqrt{4}$ (D) $\sqrt{216}$

15) Which of the below is a rational number ?

(A) $\sqrt{71}$ (B) $\sqrt{160}$

(C) $\sqrt{49}$ (D) $\sqrt{9000}$

16) Which of the below is an irrational number ?

(A) $\sqrt{2500}$ (B) $\sqrt{1600}$

(C) $\sqrt{11}$ (D) $\sqrt{121}$

17) Which of the below is an irrational number ?

(A) $\sqrt{4}$ (B) $\sqrt{45}$

(C) $\sqrt{16}$ (D) $\sqrt{100}$

18) Which of the below is an irrational number ?

(A) $\sqrt{121}$ (B) $\sqrt{6400}$

(C) $\sqrt{12100}$ (D) $\sqrt{5}$

19) Which of the below is a rational number ?

(A) $\sqrt{810}$ (B) $\sqrt{64}$

(C) $\sqrt{4000}$ (D) $\sqrt{5}$

20) Which of the below is a rational number ?

(A) $\sqrt{71}$ (B) $\sqrt{1000}$

(C) $\sqrt{9610}$ (D) $\sqrt{196}$

Grade 8

Vol 1 Week 1 Rational numbers

21) Which of the below is an irrational number ?

  (A) √2250 (B) √25

  (C) √10000 (D) √36

22) Which of the below is a rational number ?

  (A) √2560 (B) √390

  (C) √1600 (D) √2890

23) Which of the below is a rational number ?

  (A) √4410 (B) √686

  (C) √484000 (D) √840

24) Which of the below is an irrational number ?

  (A) √32 (B) √121

  (C) √4 (D) √900

25) Which of the below is an irrational number ?

  (A) √4900 (B) √1600

  (C) √225 (D) √490

26) Which of the below is a rational number ?

  (A) √25000 (B) √16000

  (C) √110 (D) √12100

27) Which of the below is a rational number ?

  (A) √40 (B) √450

  (C) √8100 (D) √100000

28) Which of the below is an irrational number ?

  (A) √800 (B) √64

  (C) √900 (D) √25

29) Which of the below is an irrational number ?

  (A) √25 (B) √6400

  (C) √40000 (D) √19

30) Which of the below is a rational number ?

  (A) √605 (B) √625

  (C) √600 (D) √117

23

Grade 8

Vol 1
Week 1
Exponents

31) Simplify to only positive exponents.

$$9^0 \cdot 9^2$$

A) $\dfrac{1}{9^3}$ B) 9^2

C) $\dfrac{1}{9^{14}}$ D) $\dfrac{1}{9^2}$

32) Simplify to only positive exponents.

$$9^6 \cdot 9^3$$

A) $\dfrac{1}{9^8}$ B) 9^{18}

C) 9^4 D) 9^9

33) Simplify to only positive exponents.

$$7 \cdot 7^9 \cdot 7^4$$

A) $\dfrac{1}{7^8}$ B) 7^{12}

C) 7^5 D) 7^{14}

34) Simplify to only positive exponents.

$$6^0 \cdot 6^5$$

A) 6^4 B) 6^{15}

C) $\dfrac{1}{6^2}$ D) 6^5

35) Simplify to only positive exponents.

$$3^2 \cdot 3^6$$

A) 3^8 B) 3^{10}

C) 3^{11} D) 3^7

36) Simplify to only positive exponents.

$$10^7 \cdot 10^{-8}$$

A) 10^4 B) 10^6

C) $\dfrac{1}{10}$ D) 10^7

37) Simplify to only positive exponents.

$$8^{-2} \cdot 8^0$$

A) 8^8 B) 8^6

C) $\dfrac{1}{8^2}$ D) 8

38) Simplify to only positive exponents.

$$2^{-7} \cdot 2^0$$

A) $\dfrac{1}{2^{16}}$ B) $\dfrac{1}{2^7}$

C) $\dfrac{1}{2^{13}}$ D) 2^3

©All rights reserved-Math-Knots LLC., VA-USA www.math-knots.com | www.a4ace.com

Grade 8

Vol 1
Week 1
Exponents

39) Simplify to only positive exponents.

$$9^8 \cdot 9^0 \cdot 9^7$$

A) 9^8 B) 9^7
C) 9^{15} D) $\dfrac{1}{9^5}$

40) Simplify to only positive exponents.

$$8^3 \cdot 8^0$$

A) 1 B) 8^3
C) $\dfrac{1}{8^{11}}$ D) 8^{16}

41) Simplify to only positive exponents.

$$4^5 \cdot 4^0$$

A) 4^4 B) 4^7
C) 4^{15} D) 4^5

42) Simplify to only positive exponents.

$$7^3 \cdot 7^8$$

A) 7^2 B) 7^{11}
C) $\dfrac{1}{7}$ D) 7^{16}

43) Simplify to only positive exponents.

$$2^{-4} \cdot 2^{10}$$

A) 2^{10} B) 2
C) 2^6 D) 2^5

44) Simplify to only positive exponents.

$$3^{10} \cdot 3^9$$

A) $\dfrac{1}{3^5}$ B) $\dfrac{1}{3^{20}}$
C) 3^{15} D) 3^{19}

45) Simplify to only positive exponents.

$$8^{-9} \cdot 8^0$$

A) 8^{17} B) $\dfrac{1}{8^9}$
C) $\dfrac{1}{8}$ D) 8^{14}

46) Simplify to only positive exponents.

$$2^2 \cdot 2^{-4}$$

A) 2^3 B) 2^{19}
C) $\dfrac{1}{2^2}$ D) 2

Grade 8

Vol 1 Week 1 Exponents

47) Simplify to only positive exponents.

$$10^{-6} \cdot 10^0$$

A) 10^4 B) 10^{12}

C) $\dfrac{1}{10^6}$ D) 10^9

48) Simplify to only positive exponents.

$$9^{-10} \cdot 9^0$$

A) $\dfrac{1}{9^6}$ B) $\dfrac{1}{9^2}$

C) $\dfrac{1}{9^{10}}$ D) 9^9

49) Simplify to only positive exponents.

$$9^{-5} \cdot 9^6$$

A) 9^{12} B) 9

C) 9^4 D) $\dfrac{1}{9^{10}}$

50) Simplify to only positive exponents.

$$3^0 \cdot 3^{-5}$$

A) $\dfrac{1}{3^5}$ B) $\dfrac{1}{3^6}$

C) 3^{19} D) $\dfrac{1}{3^2}$

51) Simplify to only positive exponents.

$$4^{-5} \cdot 4^{-3} \cdot 4^{-4}$$

A) $\dfrac{1}{4^2}$ B) 1

C) 4^{15} D) $\dfrac{1}{4^{12}}$

52) Simplify to only positive exponents.

$$6^9 \cdot 6^{-10}$$

A) $\dfrac{1}{6^3}$ B) 6^{11}

C) 6^{17} D) $\dfrac{1}{6}$

53) Simplify to only positive exponents.

$$6^0 \cdot 6^{10}$$

A) 6^{10} B) 6^2

C) 6^{12} D) $\dfrac{1}{6^{13}}$

54) Simplify to only positive exponents.

$$6^4 \cdot 6^{-1}$$

A) $\dfrac{1}{6^2}$ B) 6^8

C) 6^{13} D) 6^3

Grade 8

Vol 1 Week 1 Exponents

55) Simplify to only positive exponents.

$$9^{10} \cdot 9^8$$

A) $\dfrac{1}{9^{12}}$ B) 9^2

C) 9^7 D) 9^{18}

56) Simplify to only positive exponents.

$$4^2 \cdot 4^{-1}$$

A) 4^2 B) 4

C) 4^{12} D) 1

57) Simplify to only positive exponents.

$$8^{-2} \cdot 8^3$$

A) 8^4 B) 8^{16}

C) 1 D) 8

58) Simplify to only positive exponents.

$$3^3 \cdot 3^3$$

A) $\dfrac{1}{3^8}$ B) 3

C) 3^6 D) $\dfrac{1}{3^4}$

59) Simplify to only positive exponents.

$$3^9 \cdot 3^8$$

A) $\dfrac{1}{3^{10}}$ B) 3^{17}

C) 1 D) 3^{12}

60) Simplify to only positive exponents.

$$10 \cdot 10^{-4}$$

A) 10^2 B) $\dfrac{1}{10^3}$

C) 10^6 D) $\dfrac{1}{10^7}$

61) Simplify to only positive exponents.

$$10^6 \cdot 10^{-2}$$

A) $\dfrac{1}{10^{17}}$ B) 10^4

C) 10^5 D) 10^7

62) Simplify to only positive exponents.

$$9^7 \cdot 9^6$$

A) 9^2 B) 9^{16}

C) 9^{13} D) 9^{10}

Grade 8

Vol 1 Week 1 Exponents

63) Simplify to only positive exponents.

$$6 \cdot 6^3$$

A) 6^4 B) $\dfrac{1}{6^{15}}$

C) $\dfrac{1}{6^2}$ D) 6^{11}

64) Simplify to only positive exponents.

$$3^{-2} \cdot 3^{-7}$$

A) 3^4 B) 3

C) $\dfrac{1}{3^3}$ D) $\dfrac{1}{3^9}$

65) Simplify to only positive exponents.

$$10^7 \cdot 10^{10}$$

A) 10^{11} B) $\dfrac{1}{10^9}$

C) $\dfrac{1}{10^3}$ D) 10^{17}

66) Simplify to only positive exponents.

$$10^5 \cdot 10^0$$

A) $\dfrac{1}{10^9}$ B) $\dfrac{1}{10^3}$

C) $\dfrac{1}{10^6}$ D) 10^5

67) Simplify to only positive exponents.

$$2^3 \cdot 2^{-2}$$

A) 2 B) 2^{10}

C) $\dfrac{1}{2^{16}}$ D) 1

68) Simplify to only positive exponents.

$$2^0 \cdot 2^3$$

A) 2^{12} B) 2^9

C) 2^3 D) $\dfrac{1}{2^{11}}$

69) Simplify to only positive exponents.

$$10^2 \cdot 10^3 \cdot 10^5$$

A) 10^5 B) 10^{10}

C) 10^{14} D) $\dfrac{1}{10^{14}}$

70) Simplify to only positive exponents.

$$3^2 \cdot 3^{-1}$$

A) 3^4 B) 3^{19}

C) 3^{16} D) 3

©All rights reserved-Math-Knots LLC., VA-USA
www.math-knots.com | www.a4ace.com

Grade 8

**Vol 1
Week 1
Exponents**

71) Simplify to only positive exponents.

$$3^8 \cdot 3^2 \cdot 3^{-1}$$

A) 3^{10} B) 3^9

C) 1   D) $\dfrac{1}{3^7}$

72) Simplify to only positive exponents.

$$3^4 \cdot 3^3$$

A) 3^7 B) $\dfrac{1}{3^{11}}$

C) 3^{17} D) 3^{18}

73) Simplify to only positive exponents.

$$7^5 \cdot 7^4$$

A) 7^3 B) $\dfrac{1}{7^2}$

C) 7^6 D) 7^9

74) Simplify to only positive exponents.

$$2^{10} \cdot 2^7$$

A) $\dfrac{1}{2}$ B) 2

C) 2^{17} D) $\dfrac{1}{2^3}$

75) Simplify to only positive exponents.

$$5^0 \cdot 5^{-4}$$

A) 5^{11} B) $\dfrac{1}{5}$

C) 5^7 D) $\dfrac{1}{5^4}$

76) Simplify to only positive exponents.

$$9^5 \cdot 9^{-1}$$

A) 9^4 B) 9^{17}

C) 9^{13} D) 9^{11}

77) Simplify to only positive exponents.

$$3^2 \cdot 3^{-7} \cdot 3^5$$

A) 1   B) $\dfrac{1}{3^{10}}$

C) $\dfrac{1}{3}$ D) 3^8

78) Simplify to only positive exponents.

$$8 \cdot 8^{-8}$$

A) 8^4 B) 8^8

C) $\dfrac{1}{8^{13}}$ D) $\dfrac{1}{8^7}$

Grade 8

Vol 1 Week 1 Exponents

79) Simplify to only positive exponents.

$$3^7 \cdot 3^4$$

A) 3^9 B) 3^{10}
C) 3^{11} D) 3^{12}

80) Simplify to only positive exponents.

$$(2^{-3})^{-2}$$

A) 1 B) 2^8
C) 2^6 D) $\dfrac{1}{2^8}$

81) Simplify to only positive exponents.

$$(4^3)^3$$

A) 4^9 B) 4^4
C) $\dfrac{1}{4^8}$ D) 1

82) Simplify to only positive exponents.

$$(3^3)^0$$

A) $\dfrac{1}{3}$ B) 3^8
C) $\dfrac{1}{3^6}$ D) 1

83) Simplify to only positive exponents.

$$7^5 \cdot 7^{-8}$$

A) 7^{12} B) $\dfrac{1}{7^3}$
C) 7^5 D) 7^{10}

84) Simplify to only positive exponents.

$$3^4$$

A) 3^4 B) $\dfrac{1}{3^3}$
C) $\dfrac{1}{3^{12}}$ D) 1

85) Simplify to only positive exponents.

$$(3^3)^3$$

A) 1 B) 3^6
C) 3^9 D) 3^4

86) Simplify to only positive exponents.

$$(2^{-4})^3$$

A) $\dfrac{1}{2^{12}}$ B) 1
C) 2^6 D) $\dfrac{1}{2^4}$

Grade 8

Vol 1 Week 1 Exponents

87) Simplify to only positive exponents.

$$(2^3)^{-4}$$

A) $\dfrac{1}{2^2}$ B) 2^3

C) 1 D) $\dfrac{1}{2^{12}}$

88) Simplify to only positive exponents.

$$(4^4)^{-2}$$

A) 1 B) $\dfrac{1}{4^2}$

C) $\dfrac{1}{4^8}$ D) $\dfrac{1}{4^6}$

89) Simplify to only positive exponents.

$$(4^{-4})^4$$

A) 4^4 B) 1

C) $\dfrac{1}{4^4}$ D) $\dfrac{1}{4^{16}}$

90) Simplify to only positive exponents.

$$(3^{-2})^{-4}$$

A) 1 B) 3^8

C) $\dfrac{1}{3^4}$ D) 3^6

91) Simplify to only positive exponents.

$$(4^4)^3$$

A) 4^{12} B) 4^6

C) 4^4 D) 1

92) Simplify to only positive exponents.

$$(2^{-1})^4$$

A) 2^{12} B) 1

C) $\dfrac{1}{2^4}$ D) $\dfrac{1}{2^8}$

93) Simplify to only positive exponents.

$$(4^{-2})^3$$

A) 4^6 B) $\dfrac{1}{4^6}$

C) 4^{16} D) 1

94) Simplify to only positive exponents.

$$(2^{-3})^4$$

A) $\dfrac{1}{2^4}$ B) 1

C) 2^{12} D) $\dfrac{1}{2^{12}}$

Grade 8

Vol 1 Week 1 Exponents

95) Simplify to only positive exponents.

$$(2^3)^3$$

A) $\dfrac{1}{2^6}$ B) $\dfrac{1}{2^{12}}$

C) 2^9 D) 1

96) Simplify to only positive exponents.

$$3^{-3}$$

A) $\dfrac{1}{3^3}$ B) $\dfrac{1}{3^2}$

C) $\dfrac{1}{3^6}$ D) 1

97) Simplify to only positive exponents.

$$(4^{-3})^0$$

A) 4^4 B) 4^8

C) 4^6 D) 1

98) Simplify to only positive exponents.

$$(3^{-2})^2$$

A) $\dfrac{1}{3^4}$ B) $\dfrac{1}{3^{12}}$

C) 1   D) 3^{12}

99) Simplify to only positive exponents.

$$(4^2)^4$$

A) 4^8 B) $\dfrac{1}{4^8}$

C) $\dfrac{1}{4^3}$ D) 4^6

100) Simplify to only positive exponents.

$$(4^2)^{-3}$$

A) 4^{16} B) 1

C) $\dfrac{1}{4^6}$ D) $\dfrac{1}{4^8}$

101) Simplify to only positive exponents.

$$(2^{-1})^3$$

A) 2   B) 2^{12}

C) $\dfrac{1}{2^{12}}$ D) $\dfrac{1}{2^3}$

102) Simplify to only positive exponents.

$$(3^4)^{-2}$$

A) 1   B) $\dfrac{1}{3^6}$

C) $\dfrac{1}{3^8}$ D) 3^4

Grade 8

Vol 1
Week 1
Exponents

103) Simplify to only positive exponents.

$$(2^{-2})^0$$

A) 2^{12} B) 2^4

C) 2^6 D) 1

104) Simplify to only positive exponents.

$$(2^2)^2$$

A) $\dfrac{1}{2^2}$ B) 2^4

C) 2^8 D) 2^3

105) Simplify to only positive exponents.

$$(2^{-1})^{-1}$$

A) 1 B) $\dfrac{1}{2^8}$

C) 2 D) 2^8

106) Simplify to only positive exponents.

$$4^3$$

A) 4^3 B) 4^6

C) 1 D) $\dfrac{1}{4^3}$

107) Simplify to only positive exponents.

$$(3^{-4})^{-3}$$

A) $\dfrac{1}{3^8}$ B) 3^2

C) 3^{12} D) $\dfrac{1}{3^6}$

108) Simplify to only positive exponents.

$$(3^{-2})^{-3}$$

A) 3^{16} B) 3^6

C) 3^4 D) 1

109) Simplify to only positive exponents.

$$(2^{-2})^2$$

A) $\dfrac{1}{2^4}$ B) 2^{12}

C) $\dfrac{1}{2^9}$ D) $\dfrac{1}{2^{16}}$

110) Simplify to only positive exponents.

$$(3^{-1})^{-2}$$

A) 3^2 B) 3^8

C) $\dfrac{1}{3}$ D) 3^{12}

Grade 8

Vol 1 Week 2 Exponents

1) Simplify to only positive exponents.

$$(2^{-2})^{-3}$$

A) 2^{12} B) 1
C) $\dfrac{1}{2^{12}}$ D) 2^6

2) Simplify to only positive exponents.

$$3^3$$

A) 3^2 B) 3^3
C) 3^{12} D) $\dfrac{1}{3}$

3) Simplify to only positive exponents.

$$(2^{-2})^{-1}$$

A) 2^2 B) $\dfrac{1}{2^8}$
C) 1 D) $\dfrac{1}{2^{12}}$

4) Simplify to only positive exponents.

$$(2^3)^4$$

A) $\dfrac{1}{2^{12}}$ B) 2^{16}
C) 1 D) 2^{12}

5) Simplify to only positive exponents.

$$(4^{-1})^3$$

A) $\dfrac{1}{4^3}$ B) $\dfrac{1}{4^4}$
C) 1 D) $\dfrac{1}{4^2}$

6) Simplify to only positive exponents.

$$(4^2)^3$$

A) 4^{16} B) 4^6
C) 4^3 D) $\dfrac{1}{4^2}$

7) Simplify to only positive exponents.

$$(4^{-1})^{-3}$$

A) 4^3 B) 4^8
C) $\dfrac{1}{4^6}$ D) 4^4

8) Simplify to only positive exponents.

$$(3^0)^{-3}$$

A) $\dfrac{1}{3^{12}}$ B) 3^2
C) 1 D) 3^{12}

Grade 8

Vol 1 Week 2 Exponents

9) Simplify to only positive exponents.

$$(3^{-3})^{-2}$$

A) 3^6 B) 3^8

C) 3^2 D) $\dfrac{1}{3^{12}}$

10) Simplify to only positive exponents.

$$(4^4)^4$$

A) 4^{16} B) $\dfrac{1}{4^3}$

C) $\dfrac{1}{4^4}$ D) $\dfrac{1}{4^2}$

11) Simplify to only positive exponents.

$$(2^{-2})^3$$

A) 2^4 B) 1

C) 2^{12} D) $\dfrac{1}{2^6}$

12) Simplify to only positive exponents.

$$(4^{-3})^4$$

A) $\dfrac{1}{4^{12}}$ B) 4^6

C) 4^{12} D) $\dfrac{1}{4^4}$

13) Simplify to only positive exponents.

$$2^3$$

A) 2^3 B) 1

C) 2^4 D) 2^{12}

14) Simplify to only positive exponents.

$$3^2$$

A) 1 B) 3^2

C) $\dfrac{1}{3^{16}}$ D) 3

15) Simplify to only positive exponents.

$$(2^{-2})^4$$

A) 1 B) $\dfrac{1}{2^8}$

C) 2^8 D) 2^2

16) Simplify to only positive exponents.

$$(4^{-1})^{-2}$$

A) 4^2 B) 4^{16}

C) 4^{12} D) $\dfrac{1}{4^2}$

©All rights reserved-Math-Knots LLC., VA-USA   www.math-knots.com | www.a4ace.com

Grade 8

**Vol 1
Week 2
Exponents**

17) Simplify to only positive exponents.

$$(3^2)^{-2}$$

A) 1 B) 3^8

C) $\dfrac{1}{3^4}$ D) $\dfrac{1}{3^6}$

18) Simplify to only positive exponents.

$$(3^{-1})^4$$

A) $\dfrac{1}{3^8}$ B) 1

C) $\dfrac{1}{3^4}$ D) 3^{12}

19) Simplify to only positive exponents.

$$\dfrac{4^9}{4^7}$$

A) 4^8 B) $\dfrac{1}{4}$

C) 4^2 D) 4^9

20) Simplify to only positive exponents.

$$\dfrac{8^{-1}}{8^2}$$

A) 8^2 B) $\dfrac{1}{8^3}$

C) 8^7 D) $\dfrac{1}{8^7}$

21) Simplify to only positive exponents.

$$(3^2)^{-3}$$

A) $\dfrac{1}{3^{12}}$ B) 3^{12}

C) $\dfrac{1}{3^6}$ D) 3^9

22) Simplify to only positive exponents.

$$(3^2)^2$$

A) 3^4 B) 1

C) $\dfrac{1}{3^8}$ D) 3^2

23) Simplify to only positive exponents.

$$\dfrac{8^6}{8^9}$$

A) $\dfrac{1}{8^5}$ B) $\dfrac{1}{8^3}$

C) 8 D) 8^6

24) Simplify to only positive exponents.

$$\dfrac{2}{2^{-8}}$$

A) 2^9 B) $\dfrac{1}{2^2}$

C) $\dfrac{1}{2^3}$ D) $\dfrac{1}{2^{14}}$

©All rights reserved-Math-Knots LLC., VA-USA

Grade 8

Vol 1 Week 2 Exponents

25) Simplify to only positive exponents.

$$\frac{7^6}{7}$$

A) 7^5 B) $\frac{1}{7^3}$

C) 7^4 D) 7

26) Simplify to only positive exponents.

$$\frac{8^5}{8^3}$$

A) 8^{12} B) 8^9

C) $\frac{1}{8^{15}}$ D) 8^2

27) Simplify to only positive exponents.

$$\frac{10^4}{10^7}$$

A) 10^{14} B) $\frac{1}{10^2}$

C) $\frac{1}{10^3}$ D) 10^2

28) Simplify to only positive exponents.

$$\frac{3}{3^{-2}}$$

A) $\frac{1}{3^{16}}$ B) $\frac{1}{3^4}$

C) 3^{10} D) 3^3

29) Simplify to only positive exponents.

$$\frac{4^{-4}}{4^{-8}}$$

A) $\frac{1}{4^5}$ B) 4^4

C) $\frac{1}{4^3}$ D) 4^3

30) Simplify to only positive exponents.

$$\frac{7^3}{7^3}$$

A) 7^8 B) 1

C) $\frac{1}{7^3}$ D) 7^7

31) Simplify to only positive exponents.

$$\frac{5^{-8}}{5^9}$$

A) $\frac{1}{5^6}$ B) $\frac{1}{5^{17}}$

C) $\frac{1}{5^2}$ D) $\frac{1}{5^4}$

32) Simplify to only positive exponents.

$$\frac{10^4}{10^8}$$

A) $\frac{1}{10^4}$ B) $\frac{1}{10^9}$

C) $\frac{1}{10^8}$ D) $\frac{1}{10^{14}}$

Grade 8

Vol 1
Week 2
Exponents

33) Simplify to only positive exponents.

$$\frac{5^6}{5^{-2}}$$

A) 5^8 B) 5^2

C) $\frac{1}{5^{16}}$ D) 5^5

34) Simplify to only positive exponents.

$$\frac{9^0}{9^9}$$

A) 9^3 B) $\frac{1}{9^9}$

C) 9^7 D) $\frac{1}{9^2}$

35) Simplify to only positive exponents.

$$\frac{6^8}{6^{-1}}$$

A) 6^9 B) $\frac{1}{6^{13}}$

C) $\frac{1}{6^9}$ D) 6^{11}

36) Simplify to only positive exponents.

$$\frac{10^{10}}{10^{-10}}$$

A) 10^{20} B) 10^{10}

C) 10^7 D) 10^{17}

37) Simplify to only positive exponents.

$$\frac{3^{-4}}{3^6}$$

A) 1 B) $\frac{1}{3^{10}}$

C) 3^6 D) $\frac{1}{3^6}$

38) Simplify to only positive exponents.

$$\frac{7^{10}}{7^{-5}}$$

A) $\frac{1}{7^{11}}$ B) 7^2

C) $\frac{1}{7^2}$ D) 7^{15}

39) Simplify to only positive exponents.

$$\frac{6^9}{6^0}$$

A) $\frac{1}{6^3}$ B) $\frac{1}{6^7}$

C) 6^9 D) 1

40) Simplify to only positive exponents.

$$\frac{10^5}{10^{-9}}$$

A) 10 B) $\frac{1}{10^{11}}$

C) $\frac{1}{10^3}$ D) 10^{14}

Grade 8

Vol 1 Week 2 Exponents

41) Simplify to only positive exponents.

$$\frac{9^{-4}}{9}$$

A) 9^2 B) $\frac{1}{9^5}$

C) $\frac{1}{9}$ D) $\frac{1}{9^8}$

42) Simplify to only positive exponents.

$$\frac{5^{-8}}{5^{10}}$$

A) 5^7 B) $\frac{1}{5^7}$

C) 5^5 D) $\frac{1}{5^{18}}$

43) Simplify to only positive exponents.

$$\frac{10^2}{10^{-8}}$$

A) $\frac{1}{10}$ B) 10^{11}

C) $\frac{1}{10^2}$ D) 10^{10}

44) Simplify to only positive exponents.

$$\frac{10^9}{10^0}$$

A) 10^9 B) $\frac{1}{10^{14}}$

C) 10^8 D) 10^4

45) Simplify to only positive exponents.

$$\frac{8^2}{8^5}$$

A) 8^5 B) $\frac{1}{8^3}$

C) $\frac{1}{8}$ D) $\frac{1}{8^5}$

46) Simplify to only positive exponents.

$$\frac{4^{-8}}{4^{-4}}$$

A) $\frac{1}{4^4}$ B) 4^3

C) $\frac{1}{4^{10}}$ D) $\frac{1}{4^3}$

47) Simplify to only positive exponents.

$$\frac{5^5}{5^4}$$

A) 5 B) $\frac{1}{5^7}$

C) $\frac{1}{5^{12}}$ D) $\frac{1}{5^3}$

48) Simplify to only positive exponents.

$$\frac{4^0}{4^{-1}}$$

A) 4^7 B) 4^6

C) $\frac{1}{4^{12}}$ D) 4

Grade 8

Vol 1 Week 2 Exponents

49) Simplify to only positive exponents.

$$\frac{4^{10}}{4^{-7}}$$

A) 4^{17} B) $\frac{1}{4^8}$

C) $\frac{1}{4^2}$ D) $\frac{1}{4^3}$

50) Simplify to only positive exponents.

$$\frac{10^{-3}}{10^3}$$

A) $\frac{1}{10^5}$ B) $\frac{1}{10^7}$

C) $\frac{1}{10^6}$ D) 10^5

51) Simplify to only positive exponents.

$$\frac{3^{-3}}{3^2}$$

A) $\frac{1}{3^5}$ B) $\frac{1}{3^{13}}$

C) 3 D) $\frac{1}{3^7}$

52) Simplify to only positive exponents.

$$\frac{8^0}{8^{-4}}$$

A) 8^4 B) $\frac{1}{8^5}$

C) $\frac{1}{8^3}$ D) $\frac{1}{8^9}$

53) Simplify to only positive exponents.

$$\frac{2^4}{2^2}$$

A) 2^8 B) 2^2

C) 2 D) $\frac{1}{2^3}$

54) Simplify to only positive exponents.

$$\frac{3^3}{3^2}$$

A) $\frac{1}{3^5}$ B) 3

C) 3^2 D) $\frac{1}{3^4}$

55) Simplify to only positive exponents.

$$\frac{2^0}{2^9}$$

A) 2^6 B) 2^5

C) $\frac{1}{2^9}$ D) 2

56) Simplify to only positive exponents.

$$\frac{2^5}{2^0}$$

A) 2^5 B) 2^{18}

C) $\frac{1}{2^4}$ D) 2

©All rights reserved-Math-Knots LLC., VA-USA

www.math-knots.com | www.a4ace.com

Grade 8

Vol 1 Week 2 Exponents

57) Simplify to only positive exponents.

$$\frac{3^{-4}}{3^{-10}}$$

A) 3^8 B) $\frac{1}{3^{12}}$

C) 3^6 D) $\frac{1}{3}$

58) Simplify to only positive exponents.

$$\frac{4^7}{4^{10}}$$

A) 4^{10} B) 4^7

C) 4^8 D) $\frac{1}{4^3}$

59) Simplify to only positive exponents.

$$\frac{4^{-1}}{4^2}$$

A) 4^3 B) 4^9

C) $\frac{1}{4^3}$ D) 4^5

60) Simplify to only positive exponents.

$$\frac{4^5}{4^8}$$

A) $\frac{1}{4}$ B) $\frac{1}{4^7}$

C) $\frac{1}{4^3}$ D) 4

61) Simplify to only positive exponents.

$$\frac{10^2}{10^4}$$

A) $\frac{1}{10^9}$ B) $\frac{1}{10^2}$

C) 10^{12} D) 10^{18}

62) Simplify to only positive exponents.

$$\frac{3^5}{3^2}$$

A) 3^5 B) 3^3

C) $\frac{1}{3^{13}}$ D) $\frac{1}{3^4}$

63) Simplify to only positive exponents.

$$\frac{9^4}{9^2}$$

A) $\frac{1}{9^{10}}$ B) 9^{13}

C) $\frac{1}{9^6}$ D) 9^2

64) Simplify to only positive exponents.

$$\frac{8^6}{8^{-6}}$$

A) 8^{12} B) 8^4

C) 8^{11} D) 8

Grade 8

Vol 1 Week 2 Exponents

65) Simplify to only positive exponents.

$$\frac{4^4}{4^{-4}}$$

A) $\dfrac{1}{4^{17}}$ B) $\dfrac{1}{4^{10}}$

C) 4^8 D) $\dfrac{1}{4^4}$

66) Simplify to only positive exponents.

$$\frac{9^0}{9^{-1}}$$

A) 9^2 B) 9

C) $\dfrac{1}{9^7}$ D) $\dfrac{1}{9^5}$

67) Simplify to only positive exponents.

$$\frac{8^5}{8^5}$$

A) 8^7 B) 1

C) $\dfrac{1}{8^{20}}$ D) 8^8

68) Simplify to only positive exponents.

$$2^3 \cdot (2^2)^3$$

A) 2^{12} B) 2^{32}

C) 2^9 D) $\dfrac{1}{2^{12}}$

69) Simplify to only positive exponents.

$$\frac{9^{-5}}{9^7}$$

A) $\dfrac{1}{9^6}$ B) $\dfrac{1}{9^9}$

C) $\dfrac{1}{9^{12}}$ D) 9^5

70) Simplify to only positive exponents.

$$\frac{10^7}{10}$$

A) $\dfrac{1}{10^2}$ B) $\dfrac{1}{10^{14}}$

C) 10^2 D) 10^6

81) Simplify to only positive exponents.

$$\frac{2^4}{2^7}$$

A) $\dfrac{1}{2^3}$ B) $\dfrac{1}{2^{11}}$

C) 1 D) 2^{17}

82) Simplify to only positive exponents.

$$(2^3)^{-3} \cdot 2^2$$

A) $\dfrac{1}{2^7}$ B) 2^8

C) 2^{16} D) 1

Grade 8

Vol 1 Week 2 Exponents

73) Simplify to only positive exponents.

$$(2 \cdot (2^{-1})^0)^2$$

A) 2^2 B) 1
C) 2^9 D) 2^8

74) Simplify to only positive exponents.

$$(2^{-1})^{-1} \cdot 2^{-3}$$

A) 2^{10} B) $\dfrac{1}{2^2}$
C) 2^{12} D) 2^{21}

75) Simplify to only positive exponents.

$$2^3 \cdot ((2^2)^2 \cdot 2^{-2})^4$$

A) 2^2 B) $\dfrac{1}{2^6}$
C) 2^{11} D) 1

76) Simplify to only positive exponents.

$$2 \cdot (2^2)^4$$

A) 2^{12} B) $\dfrac{1}{2}$
C) 2^9 D) 2

77) Simplify to only positive exponents.

$$2^3 \cdot 2^{-3}$$

A) 2^{15} B) 1
C) 2^2 D) $\dfrac{1}{2^4}$

78) Simplify to only positive exponents.

$$((2^0)^0 \cdot 2^2)^0$$

A) 2^4 B) 2
C) 1 D) 2^5

79) Simplify to only positive exponents.

$$(2^3)^2 \cdot 2^3$$

A) 2^9 B) $\dfrac{1}{2^{13}}$
C) 2^5 D) $\dfrac{1}{2^3}$

80) Simplify to only positive exponents.

$$(2^4)^3 \cdot 2^0$$

A) $\dfrac{1}{2^{12}}$ B) 2^{12}
C) 1 D) $\dfrac{1}{2^{24}}$

Grade 8

Vol 1
Week 2
Exponents

81) Simplify to only positive exponents.

$$(2^{-1})^0 \cdot 2^0$$

A) $\dfrac{1}{2^{12}}$ B) 1

C) $\dfrac{1}{2^{16}}$ D) $\dfrac{1}{2^3}$

82) Simplify to only positive exponents.

$$2^3 \cdot 2^4 \cdot (2^{-3})^3$$

A) $\dfrac{1}{2^8}$ B) 2^6

C) $\dfrac{1}{2^2}$ D) $\dfrac{1}{2^{16}}$

83) Simplify to only positive exponents.

$$2^2 \cdot 2^3$$

A) 2^{10} B) 2

C) $\dfrac{1}{2^6}$ D) 2^5

84) Simplify to only positive exponents.

$$2^{-4} \cdot (2^0)^{-2}$$

A) $\dfrac{1}{2^6}$ B) $\dfrac{1}{2^{28}}$

C) 2^2 D) $\dfrac{1}{2^4}$

85) Simplify to only positive exponents.

$$(2^0)^4 \cdot 2^4$$

A) 2^2 B) 2^4

C) $\dfrac{1}{2^{28}}$ D) $\dfrac{1}{2^{10}}$

86) Simplify to only positive exponents.

$$2 \cdot (2^2)^3$$

A) 2^7 B) 2^2

C) $\dfrac{1}{2^{12}}$ D) 2

87) Simplify to only positive exponents.

$$(2^4)^3 \cdot 2^2$$

A) $\dfrac{1}{2^{12}}$ B) 2^{18}

C) 2^{14} D) $\dfrac{1}{2^3}$

88) Simplify to only positive exponents.

$$(2^{-4})^3 \cdot 2^{-1}$$

A) $\dfrac{1}{2^9}$ B) 2^6

C) $\dfrac{1}{2^{13}}$ D) 2^{18}

©All rights reserved-Math-Knots LLC., VA-USA   www.math-knots.com | www.a4ace.com

Grade 8

Vol 1 Week 2 Exponents

89) Simplify to only positive exponents.

$$2^3 \cdot (2^{-1})^0$$

A) 2^9 B) $\dfrac{1}{2^8}$

C) $\dfrac{1}{2^5}$ D) 2^3

90) Simplify to only positive exponents.

$$(2^2 \cdot 2^4)^{-1}$$

A) $\dfrac{1}{2^6}$ B) $\dfrac{1}{2^5}$

C) 2^7 D) $\dfrac{1}{2^3}$

91) Simplify to only positive exponents.

$$2^0 \cdot (2^0)^4$$

A) $\dfrac{1}{2^3}$ B) 2^4

C) 1 D) $\dfrac{1}{2^7}$

92) Simplify to only positive exponents.

$$2 \cdot (2 \cdot 2^2)^4$$

A) 2^3 B) 2^2

C) 2^{13} D) 1

93) Simplify to only positive exponents.

$$(2^0)^{-3} \cdot (2^3)^2$$

A) $\dfrac{1}{2^9}$ B) 2^2

C) 2^6 D) $\dfrac{1}{2^6}$

94) Simplify to only positive exponents.

$$2 \cdot (2^4)^3$$

A) 2^8 B) $\dfrac{1}{2^{15}}$

C) 2^{13} D) $\dfrac{1}{2}$

95) Simplify to only positive exponents.

$$(2^3)^4 \cdot 2^4$$

A) 2^{16} B) $\dfrac{1}{2^4}$

C) $\dfrac{1}{2^8}$ D) $\dfrac{1}{2^6}$

96) Simplify to only positive exponents.

$$(2^2)^2 \cdot 2^0$$

A) 1 B) 2^4

C) $\dfrac{1}{2^2}$ D) $\dfrac{1}{2^{44}}$

Grade 8

Vol 1 Week 2 Exponents

97) Simplify to only positive exponents.

$$\left(2^{-3}\right)^2 \cdot 2^0$$

A) 2^5 B) 2^{14}

C) $\dfrac{1}{2^6}$ D) $\dfrac{1}{2^7}$

98) Simplify to only positive exponents.

$$\left(2^0 \cdot 2^3\right)^0$$

A) 1 B) 2^5

C) $\dfrac{1}{2^{12}}$ D) 2^9

99) Simplify to only positive exponents.

$$\dfrac{8^{-3}}{8^2 \cdot 8^{-5}}$$

A) 8 B) $\dfrac{1}{8^6}$

C) 8^4 D) 1

100) Simplify to only positive exponents.

$$\dfrac{7^7}{7^{-3} \cdot 7^{-1}}$$

A) 7^4 B) 7^{11}

C) $\dfrac{1}{7^4}$ D) 7^9

101) Simplify to only positive exponents.

$$\left(2^3\right)^{-1} \cdot 2^4$$

A) 1 B) 2^4

C) 2 D) $\dfrac{1}{2^{14}}$

102) Simplify to only positive exponents.

$$2^4 \cdot 2^2$$

A) $\dfrac{1}{2^4}$ B) $\dfrac{1}{2^5}$

C) 2^9 D) 2^6

103) Simplify to only positive exponents.

$$\dfrac{8^8 \cdot 8^{-9}}{8^8}$$

A) 8^{22} B) 8^8

C) 8^2 D) $\dfrac{1}{8^9}$

104) Simplify to only positive exponents.

$$\dfrac{2^9 \cdot 2^4}{2^{-8}}$$

A) $\dfrac{1}{2^4}$ B) $\dfrac{1}{2^9}$

C) 2^{21} D) 2^{19}

Grade 8

Vol 1 Week 2 Exponents

105) Simplify to only positive exponents.

$$\frac{6^0}{6^4 \cdot 6^{-2}}$$

A) $\frac{1}{6^2}$ B) 6

C) 6^4 D) $\frac{1}{6^{14}}$

106) Simplify to only positive exponents.

$$\frac{6^{-2}}{6 \cdot 6^{-3}}$$

A) 1   B) 6^5

C) $\frac{1}{6^6}$ D) $\frac{1}{6^9}$

107) Simplify to only positive exponents.

$$\frac{8^8 \cdot 8^{-5} \cdot 8^3}{8^3}$$

A) 8^{11} B) 8^3

C) $\frac{1}{8^{11}}$ D) $\frac{1}{8^8}$

108) Simplify to only positive exponents.

$$\frac{7^{-6}}{7^{-3} \cdot 7^9}$$

A) $\frac{1}{7^{10}}$ B) 7^2

C) $\frac{1}{7^{12}}$ D) 1

109) Simplify to only positive exponents.

$$\frac{6^9}{6^6 \cdot 6^9}$$

A) 6^{16} B) $\frac{1}{6^6}$

C) 6^4 D) $\frac{1}{6^{10}}$

110) Simplify to only positive exponents.

$$\frac{8^4}{8 \cdot 8^5}$$

A) 8^{15} B) $\frac{1}{8^2}$

C) $\frac{1}{8^3}$ D) 8^{13}

111) Simplify to only positive exponents.

$$\frac{4^6}{4^{-7} \cdot 4^{-9} \cdot 4^3}$$

A) 4^{19} B) 4^8

C) $\frac{1}{4^4}$ D) $\frac{1}{4^{11}}$

112) Simplify to only positive exponents.

$$\frac{3^4 \cdot 3^3}{3^5}$$

A) 3^2 B) 3^4

C) $\frac{1}{3^{14}}$ D) $\frac{1}{3^{11}}$

Grade 8

Vol 1
Week 2
Exponents

113) Simplify to only positive exponents.

$$\frac{9^0}{9^{-8} \cdot 9^9}$$

A) 9^{13} B) 9^{11}

C) $\frac{1}{9^3}$ D) $\frac{1}{9}$

114) Simplify to only positive exponents.

$$\frac{9^7 \cdot 9^7}{9^2}$$

A) 9^{19} B) 9^5

C) $\frac{1}{9^4}$ D) 9^{12}

115) Simplify to only positive exponents.

$$\frac{7^8}{7^{-7} \cdot 7^{-6}}$$

A) 7^3 B) 7^{21}

C) 7^4 D) $\frac{1}{7^3}$

116) Simplify to only positive exponents.

$$\frac{9^{-2} \cdot 9^{-3}}{9^0}$$

A) $\frac{1}{9^2}$ B) 9^{16}

C) $\frac{1}{9^5}$ D) 9^{10}

117) Simplify to only positive exponents.

$$\frac{7^3 \cdot 7^{-2}}{7^8}$$

A) 1 B) 7^3

C) $\frac{1}{7^6}$ D) $\frac{1}{7^7}$

118) Simplify to only positive exponents.

$$\frac{2^7}{2^2 \cdot 2^8}$$

A) 2^9 B) 2^7

C) $\frac{1}{2^7}$ D) $\frac{1}{2^3}$

119) Simplify to only positive exponents.

$$\frac{3}{3^{-2} \cdot 3^2}$$

A) 3 B) 3^8

C) $\frac{1}{3^4}$ D) 3^2

120) Simplify to only positive exponents.

$$\frac{6^9}{6^4 \cdot 6^3}$$

A) 6^2 B) 6^5

C) 6^7 D) $\frac{1}{6^7}$

©All rights reserved-Math-Knots LLC., VA-USA

Grade 8

Vol 1 Week 3 Exponents

1) Simplify to only positive exponents.

$$\frac{5^{-1} \cdot 5^0}{5^5}$$

A) 5^{12} B) $\frac{1}{5^8}$

C) $\frac{1}{5^6}$ D) 5^7

2) Simplify to only positive exponents.

$$\frac{4^6 \cdot 4^2}{4^6}$$

A) $\frac{1}{4}$ B) $\frac{1}{4^{15}}$

C) 4^2 D) $\frac{1}{4^2}$

3) Simplify to only positive exponents.

$$\frac{8^6 \cdot 8^{-1}}{8^5}$$

A) 8^{11} B) 1

C) 8^{13} D) 8^3

4) Simplify to only positive exponents.

$$\frac{3^3 \cdot 3^0}{3^6}$$

A) $\frac{1}{3^3}$ B) 3^7

C) 3^{14} D) $\frac{1}{3^{11}}$

5) Simplify to only positive exponents.

$$\frac{9^{-4}}{9^0 \cdot 9^{-2}}$$

A) $\frac{1}{9^2}$ B) 1

C) 9^{22} D) $\frac{1}{9^{11}}$

6) Simplify to only positive exponents.

$$\frac{8^0 \cdot 8^7}{8^{-8}}$$

A) 8^8 B) $\frac{1}{8}$

C) 8^{15} D) 8^{11}

7) Simplify to only positive exponents.

$$\frac{(2^0)^{-2}}{2^3}$$

A) 2^{14} B) 2^8

C) $\frac{1}{2^4}$ D) $\frac{1}{2^3}$

8) Simplify to only positive exponents.

$$\frac{2^8}{2 \cdot 2^{-5}}$$

A) $\frac{1}{2^{15}}$ B) 2^{12}

C) 2^4 D) $\frac{1}{2^8}$

©All rights reserved-Math-Knots LLC., VA-USA

Grade 8

Vol 1 Week 3 Exponents

9) Simplify to only positive exponents.

$$\frac{7^0}{7^{-4} \cdot 7^3}$$

A) $\dfrac{1}{7^3}$ B) 7^3

C) 7 D) $\dfrac{1}{7^5}$

10) Simplify to only positive exponents.

$$\frac{2^0 \cdot 2^{-5}}{2^{-1}}$$

A) $\dfrac{1}{2^9}$ B) $\dfrac{1}{2^4}$

C) $\dfrac{1}{2^7}$ D) $\dfrac{1}{2^{12}}$

11) Simplify to only positive exponents.

$$\left(\frac{2^4}{2^0}\right)^0$$

A) 1 B) 2^2

C) $\dfrac{1}{2^{14}}$ D) 2^{57}

12) Simplify to only positive exponents.

$$\frac{2^{-3}}{(2^3)^3}$$

A) 2 B) 1

C) $\dfrac{1}{2^{14}}$ D) $\dfrac{1}{2^{12}}$

13) Simplify to only positive exponents.

$$\frac{6^2 \cdot 6^0}{6^9}$$

A) 6^{14} B) $\dfrac{1}{6^2}$

C) 6^9 D) $\dfrac{1}{6^7}$

14) Simplify to only positive exponents.

$$\frac{2^{-4}}{(2^0)^{-4}}$$

A) $\dfrac{1}{2^4}$ B) 2^4

C) 1 D) $\dfrac{1}{2^{15}}$

15) Simplify to only positive exponents.

$$\frac{2^{-2}}{2^{-1}}$$

A) 2^7 B) 2

C) $\dfrac{1}{2}$ D) $\dfrac{1}{2^2}$

16) Simplify to only positive exponents.

$$\frac{(2^4)^{-2}}{2^4}$$

A) 2^2 B) $\dfrac{1}{2^{12}}$

C) $\dfrac{1}{2^8}$ D) 2^3

Grade 8

Vol 1 Week 3 Exponents

17) Simplify to only positive exponents.

$$\frac{2^2}{(2^3)^2}$$

A) 2^8 B) $\frac{1}{2^5}$

C) $\frac{1}{2^4}$ D) 1

18) Simplify to only positive exponents.

$$\frac{(2^2)^2}{(2^3)^3}$$

A) 1 B) $\frac{1}{2^4}$

C) $\frac{1}{2^5}$ D) 2^{15}

19) Simplify to only positive exponents.

$$\frac{(2^4)^{-3}}{(2^2)^3}$$

A) $\frac{1}{2^{18}}$ B) 1

C) 2^4 D) 2^8

20) Simplify to only positive exponents.

$$\frac{2^3}{(2^0)^0}$$

A) 2^{10} B) 2^2

C) 2^3 D) $\frac{1}{2^6}$

21) Simplify to only positive exponents.

$$\frac{2}{(2^{-4})^4}$$

A) $\frac{1}{2^7}$ B) $\frac{1}{2^{21}}$

C) $\frac{1}{2^2}$ D) 2^{17}

22) Simplify to only positive exponents.

$$\frac{2^{-2}}{(2^4)^0}$$

A) $\frac{1}{2}$ B) $\frac{1}{2^2}$

C) 1 D) 2^{10}

23) Simplify to only positive exponents.

$$\left(\frac{2^{-3}}{2^2}\right)^2$$

A) $\frac{1}{2^{10}}$ B) 1

C) 2^2 D) $\frac{1}{2^4}$

24) Simplify to only positive exponents.

$$\frac{(2^4)^{-3}}{2^3}$$

A) 2^4 B) $\frac{1}{2^{10}}$

C) $\frac{1}{2^{15}}$ D) $\frac{1}{2^{12}}$

Grade 8

Vol 1
Week 3
Exponents

25) Simplify to only positive exponents.

$$\frac{2^4}{(2^{-3})^3}$$

A) 2^{13} B) $\frac{1}{2^2}$

C) $\frac{1}{2^7}$ D) 2^8

26) Simplify to only positive exponents.

$$\left(\frac{2^4}{2^0}\right)^2$$

A) 2^8 B) 2^9

C) $\frac{1}{2^{12}}$ D) 2^2

27) Simplify to only positive exponents.

$$\frac{2^{-3}}{(2^{-1})^2}$$

A) $\frac{1}{2^{13}}$ B) $\frac{1}{2}$

C) $\frac{1}{2^5}$ D) $\frac{1}{2^9}$

28) Simplify to only positive exponents.

$$\left(\frac{2}{2^{-2}}\right)^2$$

A) 2^6 B) $\frac{1}{2^2}$

C) $\frac{1}{2^{10}}$ D) 2^{16}

29) Simplify to only positive exponents.

$$\frac{(2^{-4})^4}{2^{-4}}$$

A) 2^{15} B) 1

C) $\frac{1}{2^{12}}$ D) $\frac{1}{2^6}$

30) Simplify to only positive exponents.

$$\left(\frac{2^0}{2^{-3}}\right)^3$$

A) $\frac{1}{2^2}$ B) 2^9

C) $\frac{1}{2^8}$ D) 1

31) Simplify to only positive exponents.

$$\frac{(2^2)^{-1}}{2^{-2}}$$

A) 1 B) 2^8

C) $\frac{1}{2}$ D) 2^{15}

32) Simplify to only positive exponents.

$$\frac{(2^{-4})^4}{2}$$

A) $\frac{1}{2^{11}}$ B) $\frac{1}{2^3}$

C) $\frac{1}{2^{17}}$ D) $\frac{1}{2}$

Grade 8

Vol 1 Week 3 Exponents

33) Simplify to only positive exponents.

$$\frac{(2^4)^{-4}}{2^4}$$

A) 2
B) $\frac{1}{2}$
C) $\frac{1}{2^{20}}$
D) 2^2

34) Simplify to only positive exponents.

$$\frac{(2^{-2})^{-4}}{2}$$

A) 2^7
B) 2^{12}
C) $\frac{1}{2^2}$
D) 1

35) Simplify to only positive exponents.

$$\left(\frac{2^{-1}}{2^2}\right)^4$$

A) $\frac{1}{2^7}$
B) 2^3
C) $\frac{1}{2^{12}}$
D) $\frac{1}{2}$

36) Simplify to only positive exponents.

$$\frac{2}{(2^4)^3}$$

A) 2^3
B) 2^2
C) $\frac{1}{2^{11}}$
D) $\frac{1}{2^{19}}$

37) Simplify to only positive exponents.

$$\frac{2^{-3}}{(2^4)^2}$$

A) $\frac{1}{2^6}$
B) $\frac{1}{2^{11}}$
C) 2^4
D) $\frac{1}{2}$

38) Simplify to only positive exponents.

$$\frac{(2^{-4})^3}{2^3}$$

A) $\frac{1}{2^{15}}$
B) $\frac{1}{2^3}$
C) 2^{24}
D) $\frac{1}{2^{12}}$

39) Simplify to only positive exponents.

$$\left(\frac{2^3}{2 \cdot 2^{-1}}\right)^{-1}$$

A) $\frac{1}{2^3}$
B) $\frac{1}{2^2}$
C) 2^{24}
D) $\frac{1}{2^4}$

40) Simplify to only positive exponents.

$$\frac{2}{2^{-4}}$$

A) 2
B) 2^{18}
C) 2^5
D) 2^{13}

Grade 8

Vol 1 Week 3 Exponents

41) Simplify to only positive exponents.

$$\frac{2^0}{(2^4)^0}$$

A) 1 B) 2^6

C) $\frac{1}{2^2}$ D) 2^{36}

42) Simplify to only positive exponents.

$$\frac{4}{(2^3)^2}$$

A) $\frac{1}{2^4}$ B) $\frac{1}{2^8}$

C) 2^3 D) $\frac{1}{2^5}$

43) Simplify to only positive exponents.

$$\left(\frac{2^2}{2^2 \cdot 2^3}\right)^{-3}$$

A) $\frac{1}{2^{18}}$ B) 2^9

C) $\frac{1}{2^4}$ D) 2^8

44) Simplify to only positive exponents.

$$\left(\frac{2^3}{2 \cdot 2^3}\right)^{-3}$$

A) $\frac{1}{2^5}$ B) 2^3

C) 1 D) $\frac{1}{2^6}$

45) Simplify to only positive exponents.

$$\left(\frac{2^{-2} \cdot 2^3}{2^4}\right)^3$$

A) $\frac{1}{2^{13}}$ B) $\frac{1}{2^4}$

C) $\frac{1}{2^6}$ D) $\frac{1}{2^9}$

46) Simplify to only positive exponents.

$$\frac{2^{-2} \cdot 2^{-3}}{(2^4)^4}$$

A) 2^3 B) $\frac{1}{2^9}$

C) 2^8 D) $\frac{1}{2^{21}}$

47) Simplify to only positive exponents.

$$\left(\frac{(2^0)^2}{2^2 \cdot 2^{-2}}\right)^{-3}$$

A) 2^4 B) 2^9

C) 1 D) $\frac{1}{2^{13}}$

48) Simplify to only positive exponents.

$$\frac{2^4 \cdot 2^2 \cdot 2^4}{(2^{-4})^4}$$

A) 2^8 *B) 2^{26}

C) 2^6 D) 1

©All rights reserved-Math-Knots LLC., VA-USA

Grade 8

Vol 1
Week 3
Exponents

49) Simplify to only positive exponents.

$$\frac{(2^3 \cdot 2^0)^{-3}}{2}$$

A) 2^2 B) $\frac{1}{2^{12}}$

C) $\frac{1}{2^{10}}$ D) $\frac{1}{2^{18}}$

50) Simplify to only positive exponents.

$$\frac{2^{-1} \cdot 2^2}{2^0}$$

A) $\frac{1}{2^{12}}$ B) 2

C) $\frac{1}{2^4}$ D) 2^4

51) Simplify to only positive exponents.

$$\frac{2}{(2^2)^{-4} \cdot 2^{-1}}$$

A) 2^3 B) $\frac{1}{2^{14}}$

C) 2^{10} D) 2^6

52) Simplify to only positive exponents.

$$\left(\frac{2^3}{2 \cdot 2^3 \cdot 2^{-4}}\right)^4$$

A) $\frac{1}{2^3}$ B) 2^4

C) 2^{12} D) $\frac{1}{2^{13}}$

53) Simplify to only positive exponents.

$$\frac{2^{-4} \cdot (2^0)^2}{2^4}$$

A) 2^8 B) 2^{18}

C) $\frac{1}{2^8}$ D) $\frac{1}{2^{14}}$

54) Simplify to only positive exponents.

$$\frac{2^0 \cdot 2^3}{2}$$

A) 1 B) $\frac{1}{2^9}$

C) 2^2 D) 2^6

55) Simplify to only positive exponents.

$$\frac{2^0}{(2^{-2})^3 \cdot 2^4}$$

A) 2^8 B) 2^6

C) $\frac{1}{2^3}$ D) 2^2

56) Simplify to only positive exponents.

$$\frac{2^{-1}}{(2^4)^{-2} \cdot 2^{-4}}$$

A) 2^{11} B) 2^7

C) $\frac{1}{2^2}$ D) $\frac{1}{2^{29}}$

Grade 8

Vol 1
Week 3
Exponents

57) Simplify to only positive exponents.

$$\frac{(2^{-2})^4 \cdot 2^{-3}}{2^3}$$

A) $\frac{1}{2^{14}}$ B) 2^3

C) $\frac{1}{2^2}$ D) $\frac{1}{2^8}$

58) Simplify to only positive exponents.

$$\frac{2^2 \cdot 2^3}{2^{-2}}$$

A) $\frac{1}{2}$ B) 2^7

C) 2^3 D) 2^{17}

59) Simplify to only positive exponents.

$$\left(\frac{2^2 \cdot 2^4}{2^0}\right)^3$$

A) 2^{18} B) $\frac{1}{2^3}$

C) $\frac{1}{2^6}$ D) $\frac{1}{2^9}$

60) Simplify to only positive exponents.

$$\frac{(2^4)^2}{2^2 \cdot (2^3)^4}$$

A) 2^3 B) $\frac{1}{2^6}$

C) 2^9 D) $\frac{1}{2^5}$

61) Simplify to only positive exponents.

$$\frac{(2^3)^3}{2^{-3} \cdot 2^0}$$

A) 2^3 B) $\frac{1}{2^{10}}$

C) 2^9 D) 2^{12}

62) Simplify to only positive exponents.

$$\left(\frac{2 \cdot (2^3)^2}{2^{-4}}\right)^0$$

A) $\frac{1}{2^5}$ B) 1

C) $\frac{1}{2^7}$ D) 2

63) Simplify to only positive exponents.

$$x^0 x^4$$

A) x^4 B) $6x^6$

C) $6x^5$ D) $4x^4$

64) Simplify to only positive exponents.

$$\frac{2^3 \cdot (2^3)^3}{(2^3)^0}$$

A) 2^{12} B) $\frac{1}{2^2}$

C) 2^9 D) $\frac{1}{2^4}$

Grade 8

Vol 1 Week 3 Exponents

65) Simplify to only positive exponents.

$$\frac{(2^4)^3 \cdot (2^4)^4}{2}$$

A) 2^{27} B) $\frac{1}{2}$

C) $\frac{1}{2^3}$ D) $\frac{1}{2^8}$

66) Simplify to only positive exponents.

$$b \cdot 4b^{-1}$$

A) $\frac{3}{b^3}$ B) $12b^3$

C) 4   D) $\frac{3}{b}$

67) Simplify to only positive exponents.

$$4x^3 \cdot 2x$$

A) $16x^4$ B) $9x^5$

C) $2x^6$ D) $8x^4$

68) Simplify to only positive exponents.

$$3a^4 \cdot 3a$$

A) $6a$ B) $\frac{3}{a^2}$

C) $9a^5$ D) $\frac{8}{a^3}$

69) Simplify to only positive exponents.

$$\frac{2^4}{2^4 \cdot 2^{-4}}$$

A) 2^4 B) $\frac{1}{2^8}$

C) 1   D) $\frac{1}{2^{12}}$

70) Simplify to only positive exponents.

$$2m^{-3} \cdot 4m$$

A) $\frac{6}{m^3}$ B) $4m^8$

C) $\frac{9}{m}$ D) $\frac{8}{m^2}$

71) Simplify to only positive exponents.

$$4xx^0$$

A) $\frac{6}{x^4}$ B) $8x^6$

C) $\frac{2}{x^5}$ D) $4x$

72) Simplify to only positive exponents.

$$2p^4 \cdot 4p^4$$

A) $8p^8$ B) $4p^3$

C) $6p^6$ D) $4p^4$

©All rights reserved-Math-Knots LLC., VA-USA   www.math-knots.com | www.a4ace.com

Grade 8

Vol 1 Week 3
Exponents

73) Simplify to only positive exponents.

$$2n^3 \cdot 2n^{-1}$$

A) $2n^2$ B) $4n^2$

C) $16n^8$ D) $\dfrac{4}{n^4}$

74) Simplify to only positive exponents.

$$2n^4 \cdot 4n^{-1}$$

A) $9n^4$ B) $4n^3$

C) $12n^3$ D) $8n^3$

75) Simplify to only positive exponents.

$$3k^{-1} \cdot 3k$$

A) $\dfrac{4}{k}$ B) $6k^6$

C) 9 D) 6

76) Simplify to only positive exponents.

$$a^{-2}a^2$$

A) $8a$ B) 8

C) $\dfrac{6}{a^4}$ D) 1

77) Simplify to only positive exponents.

$$x \cdot 3x^4$$

A) $3x^5$ B) $12x^4$

C) 1 D) $6x^8$

78) Simplify to only positive exponents.

$$3a \cdot 4a^3$$

A) $12a^4$ B) $8a$

C) $6a^5$ D) 12

79) Simplify to only positive exponents.

$$(v^7)^5$$

A) $\dfrac{v^{64}}{256}$ B) $\dfrac{1}{v^{90}}$

C) v^{35} D) $\dfrac{v^3}{27}$

80) Simplify to only positive exponents.

$$2x \cdot 2x^3$$

A) $8x^5$ B) $3x^8$

C) $4x^4$ D) $12x^3$

Grade 8

Vol 1 Week 3 Exponents

81) Simplify to only positive exponents.

$$n^{-4} \cdot n$$

A) $3n^5$ B) $12n^{12}$

C) $3n^8$ D) $\dfrac{1}{n^3}$

82) Simplify to only positive exponents.

$$\left(m^8\right)^{-9}$$

A) $\dfrac{1}{m^{72}}$ B) $\dfrac{1}{m^{48}}$

C) m^9 D) 8

83) Simplify to only positive exponents.

$$\left(2n^{-5}\right)^{-5}$$

A) 2187 B) n^{20}

C) $\dfrac{n^{25}}{32}$ D) $81n^4$

84) Simplify to only positive exponents.

$$\left(3p^6\right)^3$$

A) $\dfrac{243}{p^{30}}$ B) p^8

C) $\dfrac{1}{128p^{70}}$ D) $27p^{18}$

85) Simplify to only positive exponents.

$$\left(3n^2\right)^{-7}$$

A) n^{10} B) $\dfrac{1}{2187n^{14}}$

C) 1 D) $\dfrac{512}{n^{45}}$

86) Simplify to only positive exponents.

$$\left(2a^{-4}\right)^{-4}$$

A) 1 B) $\dfrac{64}{a^{12}}$

C) a^{21} D) $\dfrac{a^{16}}{16}$

87) Simplify to only positive exponents.

$$\left(2m\right)^7$$

A) $\dfrac{1}{3m^9}$ B) $\dfrac{1}{3m^4}$

C) $128m^7$ D) $\dfrac{1}{m^{24}}$

88) Simplify to only positive exponents.

$$\left(n^{-4}\right)^5$$

A) $\dfrac{1}{n^6}$ B) $\dfrac{256}{n^{72}}$

C) $\dfrac{1}{n^{20}}$ D) 1

Grade 8

Vol 1 Week 3 Exponents

89) Simplify to only positive exponents.

$$(n^4)^5$$

A) $9n^{14}$ B) $\dfrac{1}{n^{90}}$

C) n^{20} D) n^{18}

90) Simplify to only positive exponents.

$$(2r^0)^0$$

A) $9r^{16}$ B) $\dfrac{r^5}{3}$

C) $\dfrac{1}{r^{54}}$ D) 1

91) Simplify to only positive exponents.

$$(2n)^{-10}$$

A) $256n^{40}$ B) $\dfrac{1}{1024n^{10}}$

C) 1 D) 16

92) Simplify to only positive exponents.

$$(2b^4)^{-4}$$

A) $81b^4$ B) $\dfrac{1}{16b^{16}}$

C) $\dfrac{1}{256b^{64}}$ D) $243b^{35}$

93) Simplify to only positive exponents.

$$(n^9)^9$$

A) $\dfrac{1}{8}$ *B) n^{81}

C) $\dfrac{1}{n^{28}}$ D) $\dfrac{1}{8n^{21}}$

94) Simplify to only positive exponents.

$$(3x^8)^{-5}$$

A) $9x^{12}$ B) 1

C) $\dfrac{1}{x^{45}}$ D) $\dfrac{1}{243x^{40}}$

95) Simplify to only positive exponents.

$$\dfrac{5b^{10}}{6b^0}$$

A) $\dfrac{b^{17}}{4}$ B) $\dfrac{5b^{10}}{6}$

C) $\dfrac{3}{29b^{22}}$ D) $\dfrac{22}{17b^{35}}$

96) Simplify to only positive exponents.

$$(2n^{-10})^0$$

A) 1 B) n^4

C) $8n^6$ D) $\dfrac{16}{n^{28}}$

©All rights reserved-Math-Knots LLC., VA-USA   www.math-knots.com | www.a4ace.com

Grade 8

Vol 1 Week 3 Exponents

97) Simplify to only positive exponents.

$$\frac{25n^0}{23n^0}$$

A) $\dfrac{25}{23}$ B) $\dfrac{29}{25}$

C) $\dfrac{23}{9}$ D) $\dfrac{6}{n^{19}}$

98) Simplify to only positive exponents.

$$\frac{21k^{-22}}{29k^0}$$

A) $\dfrac{23}{20k^{11}}$ B) $\dfrac{8}{7k^5}$

C) $\dfrac{19}{23k^{23}}$ D) $\dfrac{21}{29k^{22}}$

99) Simplify to only positive exponents.

$$\frac{29n^{-24}}{16n^{13}}$$

A) $\dfrac{1}{8}$ B) $\dfrac{29}{16n^{37}}$

C) $\dfrac{9}{4n^{17}}$ D) $\dfrac{6}{n}$

100) Simplify to only positive exponents.

$$\frac{22m^{13}}{2m^{-27}}$$

A) $11m^{40}$ B) $\dfrac{27}{20m^7}$

C) $\dfrac{29}{12m^{31}}$ D) $\dfrac{4}{29m^{31}}$

101) Simplify to only positive exponents.

$$\frac{29v^{-10}}{22v^8}$$

A) $\dfrac{v^{14}}{13}$ B) $\dfrac{29}{22v^{18}}$

C) $\dfrac{29}{13v^9}$ D) $\dfrac{5}{4v^4}$

102) Simplify to only positive exponents.

$$\frac{7b^{29}}{28b^{-4}}$$

A) $\dfrac{b^{33}}{4}$ B) $\dfrac{9}{4b^{18}}$

C) $\dfrac{26b^7}{9}$ D) $\dfrac{13b^{34}}{20}$

103) Simplify to only positive exponents.

$$\frac{16k^{-6}}{27k^{15}}$$

A) $\dfrac{1}{k^{24}}$ B) $\dfrac{16}{27k^{21}}$

C) $\dfrac{17k^{12}}{5}$ D) $\dfrac{10k^7}{13}$

104) Simplify to only positive exponents.

$$\frac{6x^0}{17x^{-19}}$$

A) $\dfrac{16}{9x^{15}}$ B) $\dfrac{1}{2x^{48}}$

C) $\dfrac{1}{4x^{12}}$ D) $\dfrac{6x^{19}}{17}$

Grade 8

Vol 1 Week 3 Exponents

105) Simplify to only positive exponents.

$$\frac{18n^3}{19n^4}$$

A) $\frac{18}{19n}$ B) n^8

C) $\frac{7}{9n}$ D) $\frac{13}{16n^{13}}$

106) Simplify to only positive exponents.

$$\frac{12n^{-19}}{24n^0}$$

A) $\frac{1}{2n^{19}}$ B) $\frac{2}{13n^{44}}$

C) $\frac{1}{15n^{28}}$ D) $\frac{2n^{27}}{7}$

107) Simplify to only positive exponents.

$$\frac{20n^{26}}{15n^{-2}}$$

A) $\frac{4n^{28}}{3}$ B) $\frac{11n^{15}}{7}$

C) $\frac{2}{25n^{38}}$ D) $\frac{16}{15n^{20}}$

108) Simplify to only positive exponents.

$$\frac{16b^{27}}{21b^3}$$

A) $\frac{16b^{24}}{21}$ B) $\frac{5b^3}{2}$

C) $4b^{12}$ D) b^4

109) Simplify to only positive exponents.

$$\frac{v^{-11}}{25v^2}$$

A) $6v^{28}$ B) $\frac{23}{24v^5}$

C) $\frac{1}{25v^{13}}$ D) $\frac{v}{2}$

110) Simplify to only positive exponents.

$$\frac{23n^0}{20n^{-12}}$$

A) $\frac{11n^{21}}{3}$ B) $\frac{3}{7n^{12}}$

C) $\frac{3n^{13}}{5}$ D) $\frac{23n^{12}}{20}$

111) Simplify to only positive exponents.

$$\frac{2x^4}{(x^{-10})^{-7}}$$

A) $\frac{2}{x^{66}}$ B) $\frac{x^{47}}{2048}$

C) $\frac{128}{x^{42}}$ D) $\frac{2}{x^{88}}$

112) Simplify to only positive exponents.

$$(2n^{-7})^0 \cdot n^{-9}$$

A) $\frac{1}{8192n^{136}}$ B) $2n^{69}$

C) $\frac{1}{n^9}$ D) $4n^{34}$

Grade 8

Vol 1 Week 3 Exponents

113) Simplify to only positive exponents.

$$\left(r^0 \cdot r^7 \cdot 2r\right)^{13}$$

A) $8192r^{104}$ B) r^{224}

C) $\dfrac{256}{r^{16}}$ D) 1

114) Simplify to only positive exponents.

$$\left(2n^8\right)^{11} \cdot \left(n^8\right)^0$$

A) $\dfrac{1}{n^{56}}$ B) $2048n^{88}$

C) $\dfrac{n^{95}}{512}$ D) $16n^{36}$

115) Simplify to only positive exponents.

$$\left(2x^{-3}\right)^{11} \cdot x^{11}$$

A) $256x^{64}$ B) $\dfrac{8}{x^3}$

C) 16 D) $\dfrac{2048}{x^{22}}$

116) Simplify to only positive exponents.

$$\dfrac{\left(n^3\right)^6}{2n^9}$$

A) $\dfrac{n^9}{2}$ B) n^{25}

C) $\dfrac{n^{30}}{1024}$ D) $1024n^{93}$

117) Simplify to only positive exponents.

$$\left(\dfrac{2x^4}{2x^9}\right)^{-2}$$

A) $\dfrac{x^{33}}{2}$ B) x^2

C) $\dfrac{1}{2x^8}$ D) x^{10}

118) Simplify to only positive exponents.

$$\left(2n^{14} \cdot n^7\right)^8$$

A) n^6 B) 1

C) $1024n^{145}$ D) $256n^{168}$

119) Simplify to only positive exponents.

$$\dfrac{6n^5}{4n \cdot 10n^5 \cdot n^8}$$

A) $8n^7$ B) $\dfrac{5}{27n^6}$

C) $\dfrac{3}{20n^9}$ D) $\dfrac{n^{10}}{2}$

120) Simplify to only positive exponents.

$$\dfrac{\left(x^{-7}\right)^2}{2x^3}$$

A) $\dfrac{1}{2x^{17}}$ B) $\dfrac{32}{x^{50}}$

C) x^{14} D) $\dfrac{1}{x^{30}}$

Grade 8

Vol 1
Week 3
Exponents

121) Simplify to only positive exponents.

$$\frac{(2a^7)^0}{2a^8}$$

A) $\dfrac{32}{a^{14}}$ B) $256a^{25}$

C) $2a^{52}$ D) $\dfrac{1}{2a^8}$

122) Simplify to only positive exponents.

$$\frac{k^4 \cdot 2k^{-7}}{10k^7}$$

A) $\dfrac{1}{5k^{10}}$ B) $\dfrac{9k^{10}}{5}$

C) k^6 D) $\dfrac{21k^7}{4}$

123) Simplify to only positive exponents.

$$\frac{8r^7}{3r^6 \cdot 9r^9}$$

A) $\dfrac{6}{5r^2}$ B) $\dfrac{8}{27r^8}$

C) $\dfrac{1}{4r^8}$ D) $\dfrac{4}{35r^{14}}$

124) Simplify to only positive exponents.

$$\frac{5x \cdot 5x^3}{3x^7}$$

A) $\dfrac{25}{3x^3}$ B) $\dfrac{32x^2}{7}$

C) $\dfrac{81x^3}{4}$ D) $\dfrac{8}{x^{18}}$

125) Simplify to only positive exponents.

$$\frac{7x^5}{8x^0 \cdot 6x^{-6}}$$

A) $\dfrac{3x^{10}}{10}$ B) $\dfrac{7x^{11}}{48}$

C) $2x^{16}$ D) $\dfrac{80}{9x^9}$

Grade 8

1) The $\sqrt{52}$ lies between which two integers?

2) The $\sqrt{3}$ lies between which two integers?

3) The $\sqrt{37}$ lies between which two integers?

4) The $\sqrt{46}$ lies between which two integers?

5) The $\sqrt{2}$ lies between which two integers?

6) The $\sqrt{27}$ lies between which two integers?

7) The $\sqrt{65}$ lies between which two integers?

8) The $\sqrt{5}$ lies between which two integers?

9) The $\sqrt{82}$ lies between which two integers?

10) The $\sqrt{79}$ lies between which two integers?

Grade 8

Vol 1
Week 4
Square roots

11) The $\sqrt{48}$ lies between which two integers ?

12) The $\sqrt{50}$ lies between which two integers ?

13) The $\sqrt{40}$ lies between which two integers ?

14) The $\sqrt{11}$ lies between which two integers ?

15) The $\sqrt{39}$ lies between which two integers ?

16) The $\sqrt{79}$ lies between which two integers ?

17) The $\sqrt{71}$ lies between which two integers ?

18) The $\sqrt{75}$ lies between which two integers ?

19) The $\sqrt{49}$ lies between which two integers ?

20) The $\sqrt{14}$ lies between which two integers ?

Grade 8

**Vol 1
Week 4
Square roots**

21) The $\sqrt{63}$ lies between which two integers?

22) The $\sqrt{47}$ lies between which two integers?

23) The $\sqrt{93}$ lies between which two integers?

24) The $\sqrt{59}$ lies between which two integers?

25) The $\sqrt{57}$ lies between which two integers?

26) The $\sqrt{711}$ lies between which two integers?

27) The $\sqrt{798}$ lies between which two integers?

28) The $\sqrt{112}$ lies between which two integers?

29) The $\sqrt{332}$ lies between which two integers?

30) The $\sqrt{867}$ lies between which two integers?

Grade 8

**Vol 1
Week 4
Square roots**

31) The $\sqrt{634}$ lies between which two integers ?

32) The $\sqrt{287}$ lies between which two integers ?

33) The $\sqrt{865}$ lies between which two integers ?

34) The $\sqrt{109}$ lies between which two integers ?

35) The $\sqrt{784}$ lies between which two integers ?

36) The $\sqrt{828}$ lies between which two integers ?

37) The $\sqrt{197}$ lies between which two integers ?

38) The $\sqrt{509}$ lies between which two integers ?

39) The $\sqrt{672}$ lies between which two integers ?

40) The $\sqrt{341}$ lies between which two integers ?

Grade 8

**Vol 1
Week 4
Square roots**

41) $\sqrt{324}$ = ?

42) $\sqrt{361}$ = ?

43) $\sqrt{16}$ = ?

44) $-\sqrt{729}$ = ?

45) $-\sqrt{121}$ = ?

46) $-\sqrt{441}$ = ?

47) $-\sqrt{64}$ = ?

48) $\sqrt{576}$ = ?

49) $-\sqrt{196}$ = ?

50) $\sqrt{625}$ = ?

Grade 8

Vol 1
Week 4
Square roots

51) $\sqrt{64} = ?$

52) $-\sqrt{169} = ?$

53) $-\sqrt{1} = ?$

54) $-\sqrt{841} = ?$

55) $-\sqrt{4} = ?$

56) $-\sqrt{196}$

57) $-\sqrt{49} = ?$

58) $\sqrt{144} = ?$

59) $-\sqrt{529} = ?$

60) $-\sqrt{9} = ?$

Grade 8

Vol 1
Week 4
Square roots

61) $-\sqrt{400} = ?$

62) $-\sqrt{36} = ?$

63) $\sqrt{729} = ?$

64) $-\sqrt{49} = ?$

65) $-\sqrt{25} = ?$

66) $-\sqrt{\dfrac{9}{64}}$

67) $-\sqrt{\dfrac{4}{25}} = ?$

68) $-\sqrt{\dfrac{49}{64}} = ?$

69) $-\sqrt{\dfrac{81}{121}} = ?$

70) $\sqrt{\dfrac{9}{25}} = ?$

Grade 8

Vol 1
Week 4
Square roots

71) $-\sqrt{\dfrac{1}{49}} = ?$

72) $\sqrt{\dfrac{4}{36}} = ?$

73) $-\sqrt{\dfrac{4}{9}} = ?$

74) $\sqrt{\dfrac{25}{81}} = ?$

75) $-\sqrt{\dfrac{81}{100}} = ?$

76) $-\sqrt{\dfrac{1}{2500}} = ?$

77) Find the value of x for the below

$x = \sqrt{640}$

A) $10\sqrt{3}$ B) $10\sqrt{11}$
C) $8\sqrt{10}$ D) $26\sqrt{3}$

78) Find the value of x for the below

$x = \sqrt{1089}$

A) 24   B) $15\sqrt{5}$
C) 33   D) $10\sqrt{3}$

79) Find the value of x for the below

$x = \sqrt{2156}$

A) $14\sqrt{11}$ B) $7\sqrt{15}$
C) $14\sqrt{6}$ D) $28\sqrt{3}$

80) Find the value of x for the below

$x = \sqrt{896}$

A) $8\sqrt{14}$ B) $12\sqrt{13}$
C) $13\sqrt{7}$ D) $3\sqrt{15}$

81) Find the value of x for the below

$x = \sqrt{144}$

A) 12   B) $5\sqrt{7}$
C) $15\sqrt{10}$ D) $14\sqrt{10}$

Grade 8

Vol 1
Week 4
Square roots

82) Find the value of x for the below

$$x = \sqrt{72}$$

A) $12\sqrt{3}$ B) $8\sqrt{3}$
C) $6\sqrt{2}$ D) $3\sqrt{3}$

83) Find the value of x for the below

$$x = \sqrt{676}$$

A) 26   B) $14\sqrt{7}$
C) $9\sqrt{15}$ D) $3\sqrt{14}$

84) Find the value of x for the below

$$x = \sqrt{486}$$

A) $4\sqrt{6}$ B) $7\sqrt{7}$
C) $9\sqrt{6}$ D) $20\sqrt{3}$

85) Find the value of x for the below

$$x = \sqrt{450}$$

A) $15\sqrt{2}$ B) $9\sqrt{7}$
C) 30   D) $8\sqrt{2}$

86) Find the value of x for the below

$$x = \sqrt{363}$$

A) $4\sqrt{2}$ B) 16
C) 27   D) $11\sqrt{3}$

87) Find the value of x for the below

$$x = \sqrt{891}$$

A) $4\sqrt{5}$ B) $11\sqrt{14}$
C) $9\sqrt{11}$ D) $6\sqrt{15}$

88) Find the value of x for the below

$$x = \sqrt{3375}$$

A) $4\sqrt{13}$ B) $13\sqrt{6}$
C) $15\sqrt{15}$ D) $13\sqrt{11}$

89) Find the value of x for the below

$$x = \sqrt{256}$$

A) $5\sqrt{13}$ B) $9\sqrt{6}$
C) $5\sqrt{2}$ D) 16

90) Find the value of x for the below

$$x = \sqrt{252}$$

A) $10\sqrt{14}$ B) $6\sqrt{7}$
C) $15\sqrt{15}$ D) $4\sqrt{14}$

91) Find the value of x for the below

$$x = \sqrt{686}$$

A) $7\sqrt{14}$ B) $4\sqrt{13}$
C) $9\sqrt{2}$ D) 6

92) Find the value of x for the below

$$x = \sqrt{112}$$

A) $14\sqrt{2}$ B) $4\sqrt{7}$
C) $6\sqrt{14}$ D) $6\sqrt{11}$

93) Find the value of x for the below

$$x = \sqrt{2016}$$

A) $3\sqrt{11}$ B) 36
C) $12\sqrt{14}$ D) $4\sqrt{3}$

Grade 8

Vol 1 Week 4 — Square roots

94) Find the value of x for the below

$$x = \sqrt{1728}$$

A) $24\sqrt{3}$ B) 6
C) $6\sqrt{6}$ D) 16

95) Find the value of x for the below

$$x = \sqrt{180}$$

A) $13\sqrt{13}$ B) $6\sqrt{5}$
C) $14\sqrt{6}$ D) $14\sqrt{5}$

96) Find the value of x for the below

$$x = \sqrt{48}$$

A) $7\sqrt{11}$ B) $7\sqrt{7}$
C) $4\sqrt{2}$ D) $4\sqrt{3}$

97) Find the value of x for the below

$$x = \sqrt{735}$$

A) $7\sqrt{15}$ B) $12\sqrt{10}$
C) $6\sqrt{11}$ D) $2\sqrt{14}$

98) Find the value of x for the below

$$x = \sqrt{338}$$

A) $11\sqrt{14}$ B) 33
C) $13\sqrt{2}$ D) 39

99) Find the value of x for the below

$$x = \sqrt{200}$$

A) $6\sqrt{15}$ B) $10\sqrt{2}$
C) 6   D) $24\sqrt{2}$

100) Find the value of x for the below

$$x^2 = 2700$$

A) $7\sqrt{7}$ B) $11\sqrt{3}$
C) $30\sqrt{3}$ D) $7\sqrt{11}$

101) Find the value of x for the below

$$x^2 = 150$$

A) $11\sqrt{2}$ B) $5\sqrt{6}$
C) $10\sqrt{11}$ D) $7\sqrt{6}$

102) Find the value of x for the below

$$x^2 = 1134$$

A) $14\sqrt{3}$ B) $9\sqrt{13}$
C) $9\sqrt{14}$ D) $14\sqrt{15}$

103) Find the value of x for the below

$$x^2 = 245$$

A) $7\sqrt{15}$ B) $28\sqrt{2}$
C) $11\sqrt{10}$ D) $7\sqrt{5}$

104) Find the value of x for the below

$$x^2 = 900$$

A) $11\sqrt{13}$ B) $8\sqrt{3}$
C) $2\sqrt{7}$ D) 30

105) Find the value of x for the below

$$x^2 = 176$$

A) $3\sqrt{10}$ B) 16
C) $20\sqrt{2}$ D) $4\sqrt{11}$

Grade 8

Vol 1 Week 4 Square roots

106) Find the value of x for the below

$$x^2 = 99$$

A) $3\sqrt{11}$ B) $2\sqrt{2}$
C) $16\sqrt{3}$ D) $11\sqrt{15}$

107) Find the value of x for the below

$$x^2 = 360$$

A) $16\sqrt{3}$ B) $6\sqrt{14}$
C) $6\sqrt{10}$ D) 22

108) Find the value of x for the below

$$x^2 = 288$$

A) 30 B) $12\sqrt{2}$
C) $4\sqrt{3}$ D) $7\sqrt{5}$

109) Find the value of x for the below

$$x^2 = 32$$

A) 42 B) $9\sqrt{6}$
C) $14\sqrt{2}$ D) $4\sqrt{2}$

110) Find the value of x for the below

$$x^2 = 1690$$

A) $22\sqrt{3}$ B) $5\sqrt{10}$
C) $30\sqrt{3}$ D) $13\sqrt{10}$

111) Find the value of x for the below

$$x^2 = 2940$$

A) $7\sqrt{7}$ B) $9\sqrt{13}$
C) $18\sqrt{3}$ D) $14\sqrt{15}$

112) Find the value of x for the below

$$x^2 = 540$$

A) $2\sqrt{14}$ B) $6\sqrt{15}$
C) 24 D) $5\sqrt{6}$

113) Find the value of x for the below

$$x^2 = 300$$

A) $10\sqrt{10}$ B) $2\sqrt{13}$
C) $10\sqrt{3}$ D) $12\sqrt{14}$

114) Find the value of x for the below

$$x^2 = 1764$$

A) 42 B) $7\sqrt{14}$
C) $4\sqrt{10}$ D) $7\sqrt{13}$

115) Find the value of x for the below

$$x^2 = 637$$

A) $14\sqrt{13}$ B) 12
C) $7\sqrt{13}$ D) $30\sqrt{3}$

116) Find the value of x for the below

$$x^2 = 108$$

A) $6\sqrt{5}$ B) $13\sqrt{11}$
C) $26\sqrt{2}$ D) $6\sqrt{3}$

117) Find the value of x for the below

$$x^2 = 324$$

A) 18 B) $11\sqrt{15}$
C) $4\sqrt{10}$ D) $14\sqrt{6}$

Grade 8

Vol 1
Week 4
Square roots

118) Find the value of x for the below

$$x^2 = 175$$

A) $14\sqrt{10}$ B) $5\sqrt{7}$

C) $13\sqrt{11}$ D) $7\sqrt{13}$

119) Find the value of x for the below

$$x^2 = 36$$

A) 6   B) $6\sqrt{11}$

C) $14\sqrt{2}$ D) 22

120) Find the value of x for the below

$$x^2 = 567$$

A) 18   B) $11\sqrt{3}$

C) $9\sqrt{7}$ D) $13\sqrt{6}$

121) Find the value of x for the below

$$x^2 = 1815$$

A) $11\sqrt{15}$ B) 24

C) $6\sqrt{13}$ D) $15\sqrt{14}$

122) Find the value of x for the below

$$x^2 = 126$$

A) $9\sqrt{5}$ B) $15\sqrt{14}$

C) $3\sqrt{14}$ D) $14\sqrt{5}$

123) Find the value of x for the below

$$x^2 = 576$$

A) $22\sqrt{2}$ B) $5\sqrt{13}$

C) 24   D) $15\sqrt{5}$

124) Find the value of x for the below

$$x^2 = 325$$

A) $6\sqrt{3}$ B) $5\sqrt{13}$

C) $7\sqrt{3}$ D) $20\sqrt{3}$

©All rights reserved-Math-Knots LLC., VA-USA   www.math-knots.com | www.a4ace.com

Grade 8

Vol 1 Week 5 Radicals

1) Which option is equivalent to the below expression?

$$3\sqrt{3} + 2\sqrt{3}$$

A) $5\sqrt{3}$ B) $12\sqrt{3}$
C) $10\sqrt{3}$ D) $7\sqrt{3}$

2) Which option has the same value as the below expression?

$$2\sqrt{2} - 2\sqrt{2}$$

A) $4\sqrt{2}$ B) $2\sqrt{2}$
C) $6\sqrt{2}$ D) 0

3) Which option is equivalent to the below expression?

$$3\sqrt{2} + 3\sqrt{2}$$

A) $6\sqrt{2}$ B) $9\sqrt{2}$
C) $12\sqrt{2}$ D) $15\sqrt{2}$

4) Which option has the same value as the below expression?

$$-3\sqrt{2} + 2\sqrt{2}$$

A) $-\sqrt{2}$ B) $\sqrt{2}$
C) $5\sqrt{2}$ D) $3\sqrt{2}$

5) Which option is equivalent to the below expression?

$$2\sqrt{5} + 2\sqrt{5}$$

A) $8\sqrt{5}$ B) $10\sqrt{5}$
C) $4\sqrt{5}$ D) $6\sqrt{5}$

6) Which option is equivalent to the below expression?

$$-4\sqrt{6} + 4\sqrt{150}$$

A) $12\sqrt{6}$ B) $8\sqrt{6}$
C) $16\sqrt{6}$ D) $28\sqrt{6}$

7) Which option is equivalent to the below expression?

$$4\sqrt{2} + 4\sqrt{18}$$

A) $44\sqrt{2}$ B) $20\sqrt{2}$
C) $32\sqrt{2}$ D) $16\sqrt{2}$

8) Which option has the same value as the below expression?

$$-3\sqrt{3} + 3\sqrt{3}$$

A) $-3\sqrt{3}$ B) $-6\sqrt{3}$
C) 0 D) $3\sqrt{3}$

9) Which option is equivalent to the below expression?

$$3\sqrt{3} + 3\sqrt{3}$$

A) $9\sqrt{3}$ B) $15\sqrt{3}$
C) $12\sqrt{3}$ D) $6\sqrt{3}$

10) Which option has the same value as the below expression?

$$-2\sqrt{3} + 3\sqrt{3}$$

A) $\sqrt{3}$ B) $5\sqrt{3}$
C) $2\sqrt{3}$ D) $4\sqrt{3}$

©All rights reserved-Math-Knots LLC., VA-USA   www.math-knots.com | www.a4ace.com

Grade 8

Vol 1 Week 5 Radicals

11) Which option is equivalent to the below expression?

$$-3\sqrt{3} + 2\sqrt{3}$$

A) $3\sqrt{3}$ B) 0

C) $\sqrt{3}$ D) $-\sqrt{3}$

12) Which option has the same value as the below expression?

$$-2\sqrt{5} + 3\sqrt{5}$$

A) $\sqrt{5}$ B) $2\sqrt{5}$

C) 0 D) $4\sqrt{5}$

13) Which option is equivalent to the below expression?

$$-4\sqrt{5} - \sqrt{5}$$

A) $-13\sqrt{5}$ B) $-5\sqrt{5}$

C) $-17\sqrt{5}$ D) $-9\sqrt{5}$

14) Which option has the same value as the below expression?

$$-\sqrt{28} - 3\sqrt{112}$$

A) $-16\sqrt{7}$ B) $-28\sqrt{7}$

C) $-40\sqrt{7}$ D) $-14\sqrt{7}$

15) Which option is equivalent to the below expression?

$$-4\sqrt{6} + 2\sqrt{6}$$

A) $-2\sqrt{6}$ B) $4\sqrt{6}$

C) $2\sqrt{6}$ D) 0

16) Which option is equivalent to the below expression?

$$3\sqrt{8} - 3\sqrt{72}$$

A) $-18\sqrt{2}$ B) $-24\sqrt{2}$

C) $-12\sqrt{2}$ D) $-6\sqrt{2}$

17) Which option is equivalent to the below expression?

$$5\sqrt{2} - 4\sqrt{2}$$

A) $-2\sqrt{2}$ B) $\sqrt{2}$

C) $-3\sqrt{2}$ D) $-7\sqrt{2}$

18) Which option has the same value as the below expression?

$$2\sqrt{8} + 5\sqrt{2}$$

A) $21\sqrt{2}$ B) $13\sqrt{2}$

C) $9\sqrt{2}$ D) $17\sqrt{2}$

19) Which option is equivalent to the below expression?

$$-5\sqrt{5} + 2\sqrt{5}$$

A) $-4\sqrt{5}$ B) $-3\sqrt{5}$

C) $-8\sqrt{5}$ D) $-6\sqrt{5}$

20) Which option has the same value as the below expression?

$$2\sqrt{32} + 3\sqrt{2}$$

A) $30\sqrt{2}$ B) $11\sqrt{2}$

C) $22\sqrt{2}$ D) $19\sqrt{2}$

Grade 8

Vol 1 Week 5 Radicals

21) Which option is equivalent to the below expression?

$$4\sqrt{125} - 5\sqrt{5}$$

A) $15\sqrt{5}$ B) $35\sqrt{5}$
C) $25\sqrt{5}$ D) $30\sqrt{5}$

22) Which option has the same value as the below expression?

$$-4\sqrt{12} - 4\sqrt{3}$$

A) $-12\sqrt{3}$ B) $-20\sqrt{3}$
C) $-28\sqrt{3}$ D) $-24\sqrt{3}$

23) Which option is equivalent to the below expression?

$$2\sqrt{7} - 3\sqrt{7}$$

A) $-\sqrt{7}$ B) $-2\sqrt{7}$
C) 0   D) $-4\sqrt{7}$

24) Which option has the same value as the below expression?

$$5\sqrt{45} + 2\sqrt{5}$$

A) $17\sqrt{5}$ B) $49\sqrt{5}$
C) $34\sqrt{5}$ D) $32\sqrt{5}$

25) Which option is equivalent to the below expression?

$$-3\sqrt{80} - 4\sqrt{5}$$

A) $-28\sqrt{5}$ B) $-16\sqrt{5}$
C) $-20\sqrt{5}$ D) $-24\sqrt{5}$

26) Which option is equivalent to the below expression?

$$-4\sqrt{2} + 3\sqrt{8}$$

A) $-2\sqrt{2}$ B) $2\sqrt{2}$
C) $4\sqrt{2}$ D) $10\sqrt{2}$

27) Which option is equivalent to the below expression?

$$-5\sqrt{90} - 3\sqrt{250}$$

A) $-45\sqrt{10}$ B) $-30\sqrt{10}$
C) $-60\sqrt{10}$ D) $-75\sqrt{10}$

28) Which option has the same value as the below expression?

$$-2\sqrt{6} - 3\sqrt{6}$$

A) $-5\sqrt{6}$ B) $-8\sqrt{6}$
C) $-12\sqrt{6}$ D) $-10\sqrt{6}$

29) Which option is equivalent to the below expression?

$$-5\sqrt{18} - \sqrt{50}$$

A) $-40\sqrt{2}$ B) $-20\sqrt{2}$
C) $-35\sqrt{2}$ D) $-45\sqrt{2}$

30) Which option has the same value as the below expression?

$$-3\sqrt{2} - 5\sqrt{50}$$

A) $-81\sqrt{2}$ B) $-53\sqrt{2}$
C) $-78\sqrt{2}$ D) $-28\sqrt{2}$

Grade 8

Vol 1
Week 5
Radicals

31) Which option is equivalent to the below expression?

$$\sqrt{2} \cdot \sqrt{40}$$

A) $\sqrt{70}$ B) $4\sqrt{5}$
C) $\sqrt{42}$ D) 80

32) Which option has the same value as the below expression?

$$\sqrt{2} \cdot \sqrt{6}$$

A) $2\sqrt{3}$ B) $-8\sqrt{70}$
C) $2\sqrt{2}$ D) 12

33) Which option is equivalent to the below expression?

$$6\sqrt{8} \cdot 2\sqrt{8}$$

A) 8 B) 96
C) 64 D) 4

34) Which option has the same value as the below expression?

$$8\sqrt{30} \cdot \sqrt{60}$$

A) $240\sqrt{2}$ B) 1800
C) $30\sqrt{2}$ D) $3\sqrt{10}$

35) Which option is equivalent to the below expression?

$$\sqrt{5} \cdot \sqrt{5}$$

A) 5 B) $7\sqrt{6}$
C) $\sqrt{10}$ D) 25

36) Which option is equivalent to the below expression?

$$\sqrt{18} \cdot (-8\sqrt{18})$$

A) −144 B) 324
C) 6 D) 18

37) Which option is equivalent to the below expression?

$$\sqrt{90} \cdot \sqrt{10}$$

A) $\sqrt{105}$ B) 10
C) 900 D) 30

38) Which option has the same value as the below expression?

$$5\sqrt{80} - \sqrt{20}$$

A) $16\sqrt{5}$ B) $36\sqrt{5}$
C) $56\sqrt{5}$ D) $18\sqrt{5}$

39) Which option is equivalent to the below expression?

$$5\sqrt{45} - 2\sqrt{5}$$

A) $28\sqrt{5}$ B) $43\sqrt{5}$
C) $41\sqrt{5}$ D) $13\sqrt{5}$

40) Which option has the same value as the below expression?

$$(-9\sqrt{5}) \cdot \sqrt{2}$$

A) $\sqrt{7}$ B) $\sqrt{10}$
C) $-9\sqrt{10}$ D) 10

©All rights reserved-Math-Knots LLC., VA-USA

Grade 8

Vol 1 Week 5 Radicals

41) Which option is equivalent to the below expression?

$$\sqrt{63} \cdot (-10\sqrt{7})$$

A) 21 B) −210
C) 441 D) $\sqrt{70}$

42) Which option has the same value as the below expression?

$$\sqrt{30} \cdot \sqrt{70}$$

A) 10 B) 2100
C) $\sqrt{42}$ D) $10\sqrt{21}$

43) Which option is equivalent to the below expression?

$$(-6\sqrt{21}) \cdot 10\sqrt{21}$$

A) $\sqrt{42}$ B) 441
C) 21 D) −1260

44) Which option has the same value as the below expression?

$$2\sqrt{63} \cdot \sqrt{42}$$

A) 2646 B) $21\sqrt{6}$
C) $42\sqrt{6}$ D) $\sqrt{105}$

45) Which option is equivalent to the below expression?

$$\sqrt{90} \cdot \sqrt{20}$$

A) $\sqrt{110}$ B) $30\sqrt{2}$
C) 1800 D) $-6\sqrt{6}$

46) Which option is equivalent to the below expression?

$$\sqrt{24} \cdot (-2\sqrt{6})$$

A) 12 B) $\sqrt{30}$
C) 144 D) −24

47) Which option is equivalent to the below expression?

$$\sqrt{27} \cdot \sqrt{6}$$

A) $\sqrt{33}$ B) $9\sqrt{2}$
C) $\sqrt{30}$ D) 162

48) Which option has the same value as the below expression?

$$\sqrt{50} \cdot \sqrt{70}$$

A) $10\sqrt{35}$ B) $2\sqrt{30}$
C) $\sqrt{210}$ D) 3500

49) Which option is equivalent to the below expression?

$$\sqrt{10} \cdot \sqrt{50}$$

A) $\sqrt{210}$ B) $2\sqrt{15}$
C) 500 D) $10\sqrt{5}$

50) Which option has the same value as the below expression?

$$7\sqrt{70} \cdot \sqrt{28}$$

A) $98\sqrt{10}$ B) 1960
C) $14\sqrt{10}$ D) $7\sqrt{2}$

Grade 8

Vol 1 Week 5 Radicals

51) Which option is equivalent to the below expression?

$$\sqrt{50} \cdot 5\sqrt{20}$$

A) $50\sqrt{10}$ B) $10\sqrt{10}$
C) $\sqrt{70}$ D) 1000

52) Which option has the same value as the below expression?

$$\sqrt{42} \cdot \sqrt{24}$$

A) $12\sqrt{7}$ B) 1008
C) $\sqrt{30}$ D) $\sqrt{66}$

53) Which option is equivalent to the below expression?

$$\frac{2\sqrt{7}}{\sqrt{8}}$$

A) $\frac{\sqrt{14}}{2}$ B) $\frac{7\sqrt{42}}{6}$
C) $\frac{\sqrt{14}}{32}$ D) $\frac{\sqrt{10}}{5}$

54) Which option has the same value as the below expression?

$$\frac{3}{2\sqrt{6}}$$

A) $\frac{\sqrt{6}}{4}$ B) $\frac{\sqrt{14}}{7}$
C) $7\sqrt{7}$ D) $\frac{2\sqrt{6}}{3}$

55) Which option is equivalent to the below expression?

$$\sqrt{6} \cdot \sqrt{6}$$

A) 6 B) $2\sqrt{3}$
C) 36 D) $\sqrt{30}$

56) Which option is equivalent to the below expression?

$$\sqrt{60} \cdot \sqrt{24}$$

A) $2\sqrt{21}$ B) 1440
C) $6\sqrt{210}$ D) $12\sqrt{10}$

57) Which option has the same value as the below expression?

$$\sqrt{5} \cdot \sqrt{3}$$

A) $-9\sqrt{42}$ B) $\sqrt{15}$
C) $2\sqrt{2}$ D) 15

58) Which option is equivalent to the below expression?

$$2\sqrt{18} \cdot \sqrt{12}$$

A) 216 B) $6\sqrt{6}$
C) $12\sqrt{6}$ D) $\sqrt{30}$

59) Which option has the same value as the below expression?

$$10\sqrt{56} \cdot \sqrt{14}$$

A) 784 B) $\sqrt{70}$
C) 280 D) 28

Grade 8

Vol 1 Week 5 Radicals

60) Which option is equivalent to the below expression?

$$4\sqrt{30} \cdot (-9\sqrt{48})$$

A) $12\sqrt{10}$ B) $-432\sqrt{10}$

C) $\sqrt{78}$ D) 1440

61) Which option has the same value as the below expression?

$$\frac{\sqrt{2}}{\sqrt{3}}$$

A) $\frac{9\sqrt{10}}{2}$ B) $\frac{\sqrt{14}}{8}$

C) $\frac{\sqrt{6}}{3}$ D) $\frac{2\sqrt{5}}{5}$

62) Which option is equivalent to the below expression?

$$\frac{\sqrt{10}}{\sqrt{6}}$$

A) $\frac{\sqrt{15}}{3}$ B) $\frac{\sqrt{15}}{5}$

C) $\frac{3\sqrt{70}}{7}$ D) $\frac{9\sqrt{42}}{14}$

63) Which option has the same value as the below expression?

$$\frac{\sqrt{9}}{2\sqrt{45}}$$

A) $2\sqrt{5}$ B) $\frac{7\sqrt{14}}{4}$

C) $-\frac{\sqrt{10}}{18}$ D) $\frac{\sqrt{5}}{10}$

64) Which option is equivalent to the below expression?

$$\frac{\sqrt{8}}{5\sqrt{3}}$$

A) $\frac{5\sqrt{6}}{4}$ B) $\frac{\sqrt{2}}{10}$

C) $\frac{2\sqrt{6}}{15}$ D) $\frac{\sqrt{6}}{3}$

65) Which option has the same value as the below expression?

$$\frac{7}{9\sqrt{8}}$$

A) $\frac{7\sqrt{2}}{36}$ B) $\frac{3\sqrt{5}}{5}$

C) $\frac{7\sqrt{2}}{40}$ D) $\frac{18\sqrt{2}}{7}$

Grade 8

Vol 1 Week 5 Radicals

66) Which option is equivalent to the below expression?

$$\frac{2\sqrt{2}}{8\sqrt{10}}$$

A) $4\sqrt{5}$ B) $\frac{\sqrt{5}}{20}$

C) $\frac{\sqrt{6}}{4}$ D) $\frac{2\sqrt{14}}{7}$

67) Which option has the same value as the below expression?

$$\frac{\sqrt{40}}{\sqrt{28}}$$

A) $\frac{\sqrt{30}}{30}$ B) $\frac{\sqrt{70}}{10}$

C) $\frac{18\sqrt{5}}{5}$ D) $\frac{\sqrt{70}}{7}$

68) Which option is equivalent to the below expression?

$$\frac{5\sqrt{6}}{\sqrt{7}}$$

A) $2\sqrt{10}$ B) $\frac{\sqrt{30}}{5}$

C) $\frac{5\sqrt{42}}{}$ D) $\sqrt{42}$

69) Which option is equivalent to the below expression?

$$\frac{8\sqrt{21}}{\sqrt{35}}$$

A) $\frac{\sqrt{15}}{24}$ B) $\frac{4\sqrt{14}}{7}$

C) $\frac{\sqrt{6}}{6}$ D) $\frac{8\sqrt{15}}{5}$

70) Which option is equivalent to the below expression?

$$\frac{9\sqrt{18}}{7\sqrt{60}}$$

A) $\frac{9\sqrt{30}}{70}$ B) $\frac{10\sqrt{15}}{27}$

C) $\frac{\sqrt{35}}{7}$ D) $\frac{7\sqrt{30}}{27}$

71) Which option has the same value as the below expression?

$$\frac{\sqrt{70}}{\sqrt{60}}$$

A) $2\sqrt{3}$ B) $\frac{\sqrt{42}}{6}$

C) $\frac{\sqrt{70}}{7}$ D) $\frac{\sqrt{42}}{7}$

Grade 8

Vol 1
Week 5
Radicals

72) Which option is equivalent to the below expression?

$$\frac{7\sqrt{10}}{3\sqrt{6}}$$

A) $\dfrac{7\sqrt{15}}{9}$ B) $\dfrac{\sqrt{70}}{90}$

C) $\dfrac{27\sqrt{2}}{4}$ D) $\sqrt{2}$

73) Which option has the same value as the below expression?

$$\frac{5\sqrt{4}}{9\sqrt{6}}$$

A) $\dfrac{9\sqrt{6}}{10}$ B) $\dfrac{5\sqrt{6}}{27}$

C) $\dfrac{3\sqrt{10}}{8}$ D) $\dfrac{\sqrt{6}}{24}$

74) Which option is equivalent to the below expression?

$$\frac{\sqrt{8}}{8\sqrt{3}}$$

A) $\dfrac{5\sqrt{6}}{6}$ B) $\dfrac{\sqrt{42}}{14}$

C) $2\sqrt{6}$ D) $\dfrac{\sqrt{6}}{12}$

75) Which option is equivalent to the below expression?

$$\frac{3\sqrt{45}}{7\sqrt{18}}$$

A) $\dfrac{7\sqrt{10}}{15}$ B) $\dfrac{10\sqrt{14}}{7}$

C) $\dfrac{3\sqrt{10}}{14}$ D) $-\dfrac{\sqrt{5}}{20}$

76) Which option is equivalent to the below expression?

$$\frac{8\sqrt{3}}{\sqrt{10}}$$

A) $\dfrac{4\sqrt{30}}{5}$ B) $\dfrac{\sqrt{6}}{15}$

C) $\dfrac{\sqrt{30}}{15}$ D) $\dfrac{3\sqrt{5}}{10}$

77) Which option has the same value as the below expression?

$$\frac{9}{\sqrt{6}}$$

A) $\dfrac{2\sqrt{7}}{49}$ B) $\dfrac{\sqrt{6}}{9}$

C) $\dfrac{\sqrt{70}}{5}$ D) $\dfrac{3\sqrt{6}}{2}$

Grade 8

Vol 1 Week 5 Radicals

78) Which option has the same value as the below expression?

$$\frac{9\sqrt{9}}{5\sqrt{8}}$$

A) $\dfrac{9\sqrt{3}}{4}$ B) $\dfrac{5\sqrt{21}}{63}$

C) $\dfrac{27\sqrt{2}}{20}$ D) $\dfrac{\sqrt{15}}{10}$

79) Which option is equivalent to the below expression?

$$\frac{3\sqrt{64}}{\sqrt{40}}$$

A) $\dfrac{6\sqrt{10}}{5}$ B) $\dfrac{\sqrt{10}}{14}$

C) $\sqrt{3}$ D) $\dfrac{5\sqrt{10}}{2}$

80) Which option is equivalent to the below expression?

$$\frac{\sqrt{36}}{10\sqrt{42}}$$

A) $\dfrac{3\sqrt{10}}{10}$ B) $\dfrac{5\sqrt{42}}{3}$

C) $\dfrac{\sqrt{10}}{10}$ D) $\dfrac{\sqrt{42}}{70}$

81) Which option is equivalent to the below expression?

$$\frac{\sqrt{9}}{\sqrt{6}}$$

A) $\dfrac{\sqrt{14}}{20}$ B) $\dfrac{\sqrt{6}}{2}$

C) $\dfrac{\sqrt{6}}{3}$ D) $\dfrac{3\sqrt{6}}{2}$

82) Simplify the below

$$\sqrt[3]{24x^3}$$

A) $5\sqrt[3]{5x}$ B) $3x\sqrt[3]{4x^2}$

C) $5x\sqrt[3]{2x}$ *D) $2x\sqrt[3]{3}$

83) Simplify the below

$$\sqrt[3]{-128x^5}$$

A) $-4x\sqrt[3]{2x^2}$ B) $-2x\sqrt[3]{5x^2}$

C) $4x\sqrt[3]{3x}$ D) $5x\sqrt[3]{3x}$

84) Simplify the below

$$\sqrt[5]{-64n}$$

A) $2n\sqrt[5]{5n}$ B) $2\sqrt[5]{3n}$

C) $-2\sqrt[5]{2n}$ D) $-2\sqrt[5]{2n^3}$

85) Simplify the below

$$\sqrt[3]{16n}$$

A) $2n\sqrt[3]{3n^2}$ B) $2\sqrt[3]{2n}$

C) $4n\sqrt[3]{5n^2}$ D) $3\sqrt[3]{3n}$

Grade 8

**Vol 1
Week 5
Radicals**

86) Simplify the below

$$\sqrt[4]{112}$$

A) $2\sqrt[4]{8}$ B) $2\sqrt[4]{7}$

C) $2\sqrt[4]{3}$ D) $2\sqrt[4]{5}$

87) Simplify the below

$$\sqrt[3]{24}$$

A) $4\sqrt[3]{4}$ B) $-2\sqrt[3]{6}$

C) $5\sqrt[3]{3}$ D) $2\sqrt[3]{3}$

88) Simplify the below

$$\sqrt[7]{-1280}$$

A) $2\sqrt[7]{5}$ B) $2\sqrt[7]{6}$

C) $2\sqrt[7]{10}$ D) $-2\sqrt[7]{10}$

89) Simplify the below

$$\sqrt[4]{648}$$

A) $3\sqrt[4]{8}$ B) $3\sqrt[4]{2}$

C) $3\sqrt[4]{9}$ D) $2\sqrt[4]{10}$

90) Simplify the below

$$\sqrt[4]{32}$$

A) $2\sqrt[4]{5}$ B) $3\sqrt[4]{3}$

C) $2\sqrt[4]{2}$ D) $3\sqrt[4]{8}$

91) Simplify the below

$$\sqrt[3]{80}$$

A) $2\sqrt[3]{10}$ B) 8

C) $4\sqrt[3]{2}$ D) $4\sqrt[3]{10}$

92) Simplify the below

$$\sqrt[3]{375}$$

A) $5\sqrt[3]{3}$ B) $-2\sqrt[3]{6}$

C) $-5\sqrt[3]{5}$ D) $4\sqrt[3]{4}$

93) Simplify the below

$$\sqrt[3]{-640}$$

A) $-4\sqrt[3]{10}$ B) $4\sqrt[3]{6}$

C) $5\sqrt[3]{6}$ D) $5\sqrt[3]{4}$

94) Simplify the below

$$\sqrt[4]{810}$$

A) $3\sqrt[4]{10}$ B) $2\sqrt[4]{10}$

C) $2\sqrt[4]{3}$ D) $3\sqrt[4]{8}$

95) Simplify the below

$$\sqrt[3]{128}$$

A) $3\sqrt[3]{3}$ B) $-2\sqrt[3]{10}$

C) $-3\sqrt[3]{5}$ D) $4\sqrt[3]{2}$

Grade 8

Vol 1 Week 5 Radicals

96) Simplify the below

$$\sqrt[3]{-750}$$

A) $2\sqrt[3]{10}$ B) $-4\sqrt[3]{4}$
C) $-5\sqrt[3]{6}$ D) $5\sqrt[3]{9}$

97) Simplify the below

$$\sqrt[3]{625}$$

A) $5\sqrt[3]{5}$ B) $3\sqrt[3]{7}$
C) $4\sqrt[3]{5}$ D) $2\sqrt[3]{4}$

98) Simplify the below

$$\sqrt[4]{567}$$

A) $2\sqrt[4]{6}$ B) $3\sqrt[4]{6}$
C) $3\sqrt[4]{7}$ D) $3\sqrt[4]{9}$

99) Simplify the below

$$\sqrt[3]{-40}$$

A) $-3\sqrt[3]{7}$ B) $-4\sqrt[3]{4}$
C) $3\sqrt[3]{2}$ D) $-2\sqrt[3]{5}$

100) Simplify the below

$$\sqrt[3]{-1125}$$

A) $-5\sqrt[3]{9}$ B) $4\sqrt[3]{10}$
C) $5\sqrt[3]{6}$ D) $4\sqrt[3]{7}$

101) Simplify the below

$$\sqrt[3]{32}$$

A) $2\sqrt[3]{4}$ B) $2\sqrt[3]{9}$
C) $3\sqrt[3]{4}$ D) $2\sqrt[3]{3}$

102) Simplify the below

$$\sqrt[6]{128}$$

A) $2\sqrt[6]{7}$ B) $2\sqrt[6]{2}$
C) $2\sqrt[6]{4}$ D) $2\sqrt[6]{5}$

103) Simplify the below

$$\sqrt[3]{250}$$

A) $4\sqrt[3]{2}$ B) $4\sqrt[3]{9}$
C) $5\sqrt[3]{2}$ D) $5\sqrt[3]{7}$

104) Simplify the below

$$\sqrt[3]{320}$$

A) $2\sqrt[3]{6}$ B) $-4\sqrt[3]{9}$
C) $-5\sqrt[3]{9}$ D) $4\sqrt[3]{5}$

105) Simplify the below

$$\sqrt[7]{-512}$$

A) $-2\sqrt[7]{4}$ B) $-2\sqrt[7]{9}$
C) $2\sqrt[7]{10}$ D) $-2\sqrt[7]{2}$

©All rights reserved-Math-Knots LLC., VA-USA

www.math-knots.com | www.a4ace.com

Grade 8

Vol 1
Week 5
Radicals

106) Simplify the below

$$\sqrt[7]{256}$$

A) $2\sqrt[7]{2}$ B) $2\sqrt[7]{7}$

C) $-2\sqrt[7]{9}$ D) $2\sqrt[7]{5}$

107) Simplify the below

$$\sqrt[4]{80}$$

A) $3\sqrt[4]{4}$ B) $3\sqrt[4]{6}$

C) $2\sqrt[4]{5}$ D) $3\sqrt[4]{10}$

108) Simplify the below

$$\sqrt[4]{96}$$

A) $3\sqrt[4]{5}$ B) $2\sqrt[4]{6}$

C) $2\sqrt[4]{9}$ D) $3\sqrt[4]{3}$

109) Simplify the below

$$\sqrt[3]{-625}$$

A) $-5\sqrt[3]{5}$ B) $4\sqrt[3]{3}$

C) $3\sqrt[3]{5}$ D) $4\sqrt[3]{10}$

110) Simplify the below

$$\sqrt[3]{81}$$

A) $-5\sqrt[3]{7}$ B) $3\sqrt[3]{3}$

C) $2\sqrt[3]{7}$ D) $-2\sqrt[3]{5}$

111) Simplify the below

$$\sqrt[3]{-448}$$

A) $-4\sqrt[3]{7}$ B) $3\sqrt[3]{2}$

C) -10 D) 8

112) Simplify the below

$$\sqrt[3]{448}$$

A) $4\sqrt[3]{7}$ B) $5\sqrt[3]{10}$

C) $4\sqrt[3]{4}$ D) 6

113) Simplify the below

$$\sqrt[3]{-375}$$

A) $-5\sqrt[3]{3}$ B) $3\sqrt[3]{2}$

C) $-2\sqrt[3]{6}$ D) 10

114) Simplify the below

$$\sqrt[3]{72}$$

A) $2\sqrt[3]{9}$ B) $3\sqrt[3]{5}$

C) $-4\sqrt[3]{7}$ D) $4\sqrt[3]{4}$

115) Simplify the below

$$\sqrt[3]{1250}$$

A) $5\sqrt[3]{10}$ B) $-2\sqrt[3]{9}$

C) 8 D) $-2\sqrt[3]{7}$

©All rights reserved-Math-Knots LLC., VA-USA

Grade 8

Vol 1
Week 5
Radicals

116) Simplify the below

$$\sqrt[3]{64}$$

A) $5\sqrt[3]{10}$ B) -10

C) $2\sqrt[3]{9}$ D) 4

117) Simplify the below

$$\sqrt[3]{216}$$

A) 6 B) $-4\sqrt[3]{3}$

C) $5\sqrt[3]{9}$ D) $3\sqrt[3]{10}$

118) Simplify the below

$$\sqrt[4]{405}$$

A) $3\sqrt[4]{6}$ B) $3\sqrt[4]{3}$

C) $3\sqrt[4]{5}$ D) $3\sqrt[4]{2}$

119) Simplify the below

$$\sqrt[4]{160}$$

A) $2\sqrt[4]{10}$ B) $3\sqrt[4]{9}$

C) $2\sqrt[4]{6}$ D) $3\sqrt[4]{2}$

120) Simplify the below

$$\sqrt[3]{1000}$$

A) $5\sqrt[3]{6}$ B) 10

C) $-3\sqrt[3]{9}$ D) $2\sqrt[3]{7}$

121) Simplify the below

$$\sqrt[3]{40n^4}$$

A) $4n\sqrt[3]{3}$ B) $2n\sqrt[3]{5n}$

C) $-5n\sqrt[3]{2}$ D) $-2n\sqrt[3]{4n^2}$

122) Simplify the below

$$\sqrt[3]{192x^2}$$

A) $2x\sqrt[3]{3}$ B) $-5\sqrt[3]{3x}$

C) $4\sqrt[3]{3x^2}$ D) $5\sqrt[3]{5x^2}$

123) Simplify the below

$$\sqrt[5]{64m}$$

A) $2\sqrt[5]{2m^2}$ B) $2\sqrt[5]{3m^4}$

C) $2\sqrt[5]{2m}$ D) $2\sqrt[5]{5m^4}$

124) Simplify the below

$$\sqrt[3]{256}$$

A) $4\sqrt[3]{4}$ B) $2\sqrt[3]{10}$

C) 6 D) $3\sqrt[3]{4}$

125) Simplify the below

$$\sqrt[3]{40}$$

A) $4\sqrt[3]{5}$ B) $4\sqrt[3]{10}$

C) $2\sqrt[3]{5}$ D) 10

Grade 8

**Vol 1
Week 5
Radicals**

126) Simplify the below

$$\sqrt[3]{875}$$

A) $2\sqrt[3]{2}$ B) $3\sqrt[3]{10}$

C) $5\sqrt[3]{7}$ D) $-3\sqrt[3]{3}$

127) Simplify the below

$$\sqrt[3]{1125}$$

A) $2\sqrt[3]{2}$ B) $-2\sqrt[3]{2}$

C) $4\sqrt[3]{10}$ D) $5\sqrt[3]{9}$

128) Simplify the below

$$\sqrt[6]{448}$$

A) $2\sqrt[6]{7}$ B) $2\sqrt[6]{4}$

C) $2\sqrt[6]{3}$ D) $2\sqrt[6]{2}$

129) Simplify the below

$$\sqrt[5]{64m^3}$$

A) $2\sqrt[5]{2m^3}$ B) $2m\sqrt[5]{3}$

C) $-2m\sqrt[5]{5m}$ D) $-2m\sqrt[5]{4m^2}$

130) Simplify the below

$$\sqrt[5]{-96p^5}$$

A) $-2p\sqrt[5]{3}$ B) $2\sqrt[5]{5p^4}$

C) $2\sqrt[5]{2p^2}$ D) $2\sqrt[5]{2p^3}$

131) Simplify the below

$$\sqrt[3]{500x^3}$$

A) $2x\sqrt[3]{3}$ B) $2x\sqrt[3]{2}$

C) $4x\sqrt[3]{3x}$ D) $5x\sqrt[3]{4}$

132) Simplify the below

$$\sqrt[7]{512x^8}$$

A) $2\sqrt[7]{5x^5}$ B) $2x\sqrt[7]{4x}$

C) $2x\sqrt[7]{3x}$ D) $2x\sqrt[7]{4}$

133) Simplify the below

$$\sqrt[5]{-128m^4}$$

A) $2\sqrt[5]{3m}$ B) $-2\sqrt[5]{4m^4}$

C) $2\sqrt[5]{2m^3}$ D) $2m\sqrt[5]{3m}$

134) Simplify the below

$$\sqrt[3]{625x^5}$$

A) $2x\sqrt[3]{2x}$ B) $5x\sqrt[3]{5x^2}$

C) $5\sqrt[3]{3x^2}$ D) $-2\sqrt[3]{5x}$

135) Simplify the below

$$\sqrt[3]{128n^4}$$

A) $4\sqrt[3]{5n}$ B) $2n\sqrt[3]{2n}$

C) $3\sqrt[3]{2n^2}$ D) $4n\sqrt[3]{2n}$

Grade 8

Vol 1 — Week 6 — Scientific Notation

1) Which of the below represents the standard notation of
$$9.42 \times 10^5$$

A) 94200000 B) 0.00942

C) 9420000 D) 942000

2) Which of the below represents the standard notation of
$$8.8 \times 10^5$$

A) 8.8 B) 880000

C) 8800000 D) 88000000

3) Which of the below represents the standard notation of
$$6 \times 10^3$$

A) 600 B) 0.006

C) 6000 D) 6

4) Which of the below represents the standard notation of
$$3.4 \times 10^{-4}$$

A) 0.0034 B) 3400

C) 34000 D) 0.00034

5) Which of the below represents the standard notation of
$$7 \times 10^2$$

A) 7000 B) 70

C) 700 D) 7

6) Which of the below represents the standard notation of
$$3.6 \times 10^6$$

A) 36000000 B) 0.00036

C) 3600000 D) 0.000036

7) Which of the below represents the standard notation of
$$9.44 \times 10^{-3}$$

A) 944000 B) 0.0944

C) 0.00944 D) 0.944

8) Which of the below represents the standard notation of
$$8 \times 10^7$$

A) 8 B) 80000000

C) 80 D) 0.8

9) Which of the below represents the standard notation of
$$9.9 \times 10^8$$

A) 0.0000000099 B) 990000000

C) 9900000000 D) 0.000000099

10) Which of the below represents the standard notation of
$$8.1 \times 10^{-6}$$

A) 0.0081 B) 0.0000081

C) 0.000081 D) 0.00081

Grade 8

Vol 1 Week 6 Scientific Notation

11) Which of the below represents the standard notation of

$$9.49 \times 10^{-7}$$

A) 0.0000000095 B) 0.000000949

C) 0.0000000949 D) 0.0000000009

12) Which of the below represents the standard notation of

$$7.2 \times 10^{-8}$$

A) 0.000000072 B) 0.0000072

C) 0.00000072 D) 0.072

13) Which of the below represents the standard notation of

$$3 \times 10^{-1}$$

A) 0.3 B) 0.0003

C) 0.003 D) 0.03

14) Which of the below represents the standard notation of

$$3 \times 10^{3}$$

A) 3000 B) 0.00003

C) 0.0003 D) 0.003

15) Which of the below represents the standard notation of

$$2.79 \times 10^{2}$$

A) 0.0279 B) 2790

C) 279 D) 0.279

16) Which of the below represents the standard notation of

$$3 \times 10^{-3}$$

A) 0.003 B) 30000

C) 0.00003 D) 0.0003

17) Which of the below represents the standard notation of

$$8.61 \times 10^{7}$$

A) 0.0000000861 B) 0.000000861

C) 8610000 D) 86100000

18) Which of the below represents the standard notation of

$$2.4 \times 10^{-6}$$

A) 0.00000024 B) 240000000

C) 0.0000024 D) 24000000

19) Which of the below represents the standard notation of

$$5.4 \times 10^{-5}$$

A) 0.000054 B) 0.0054

C) 540000 D) 0.00054

20) Which of the below represents the standard notation of

$$8.9 \times 10^{5}$$

A) 0.000089 B) 0.0000089

C) 0.00000089 D) 890000

Grade 8

Vol 1 Week 6 — Scientific Notation

21) Which of the below represents the standard notation of

$$4 \times 10^9$$

A) 0.0000004 B) 4000000000
C) 0.00000004 D) 0.000004

22) Which of the below represents the standard notation of

$$2.9 \times 10^{-1}$$

A) 0.029 B) 0.0029
C) 0.29 D) 0.00029

23) Which of the below represents the standard notation of

$$5.3 \times 10^{-7}$$

A) 0.000053 B) 53000000
C) 0.0000053 D) 0.00000053

24) Which of the below represents the standard notation of

$$4.83 \times 10^7$$

A) 48300000 B) 4830000
C) 4.83 D) 48.3

25) Which of the below represents the standard notation of

$$6.7 \times 10^5$$

A) 670000 B) 67000
C) 6700 D) 0.00000067

26) Which of the below represents the standard notation of

$$3 \times 10^{-10}$$

A) 0.003 B) 0.0000000003
C) 3000 D) 30000

27) Which of the below represents the standard notation of

$$6.9 \times 10^9$$

A) 0.0069 B) 0.069
C) 6.9 D) 6900000000

28) Which of the below represents the standard notation of

$$3.6 \times 10^{-10}$$

A) 3.6 B) 36
C) 0.00000000036 D) 0.36

29) Which of the below represents the standard notation of

$$6 \times 10^{-8}$$

A) 6000000 B) 0.0000006
C) 0.00000006 D) 60000000

30) Which of the below represents the standard notation of

$$6 \times 10^4$$

A) 6000 B) 0.06
C) 0.006 D) 60000

Grade 8

Vol 1 Week 6 Scientific Notation

31) Which of the below represents the standard notation of
$$4.09 \times 10^{-1}$$
A) 0.409 B) 40900
C) 4090 D) 0.0409

32) Which of the below represents the standard notation of
$$8.4 \times 10^{-5}$$
A) 840000000 B) 0.00000084
C) 0.0000084 D) 0.000084

33) Which of the below represents the standard notation of
$$3.95 \times 10^{0}$$
A) 0.00395 B) 0.0395
C) 0.395 D) 3.95

34) Which of the below represents the standard notation of
$$9.3 \times 10^{8}$$
A) 930000000 B) 0.0000093
C) 0.00000093 D) 93000000

35) Which of the below represents the standard notation of
$$2.7 \times 10^{1}$$
A) 0.27 B) 2.7
C) 270 D) 27

36) Which of the below represents the standard notation of
$$8.4 \times 10^{4}$$
A) 0.0000084 B) 0.000084
C) 0.00084 D) 84000

37) Which of the below represents the standard notation of
$$7.3 \times 10^{4}$$
A) 73000 B) 0.0073
C) 0.073 D) 0.00073

38) Which of the below represents the standard notation of
$$9.99 \times 10^{5}$$
A) 0.00999 B) 999000
C) 0.000999 D) 0.0000999

39) Which of the below represents the standard notation of
$$8.75 \times 10^{4}$$
A) 0.0000875 B) 87500
C) 8750 D) 875

40) Which of the below represents the standard notation of
$$2.9 \times 10^{6}$$
A) 0.000029 B) 0.0000029
C) 0.00000029 D) 2900000

Grade 8

Vol 1 Week 6 Scientific Notation

41) Which of the below represents the standard notation of
$$2.61 \times 10^2$$

A) 26.1 B) 0.261
C) 261 D) 0.0261

42) Which of the below represents the standard notation of
$$7 \times 10^{10}$$

A) 700000000000 B) 70000000000
C) 700000000 D) 7000000000

43) Which of the below represents the standard notation of
$$9.2 \times 10^4$$

A) 0.00092 B) 920000
C) 920 D) 92000

44) Which of the below represents the standard notation of
$$6.4 \times 10^3$$

A) 640 B) 64
C) 6400 D) 0.0064

45) Which of the below represents the standard notation of
$$1.8 \times 10^{-2}$$

A) 0.018 B) 0.18
C) 18 D) 1.8

46) Which of the below represents the standard notation of
$$5 \times 10^{-8}$$

A) 0.00000005 B) 0.000000005
C) 0.0000000005 D) 5000000

47) Which of the below represents the standard notation of
$$7.7 \times 10^{10}$$

A) 0.00077 B) 0.000077
C) 77000000000 D) 770000000000

48) Which of the below represents the standard notation of
$$9.8 \times 10^0$$

A) 9.8 B) 0.098
C) 0.98 D) 98

49) Which of the below represents the standard notation of
$$9.8 \times 10^4$$

A) 0.00098 B) 980000
C) 98000 D) 9800000

50) Which of the below represents the standard notation of
$$2.7 \times 10^3$$

A) 2700 B) 0.027
C) 27000 D) 270

Grade 8

Vol 1 — Week 6 — Scientific Notation

51) Which of the below represents the scientific notation of

7200

A) 0.72×10^3 B) 7.2×10^3

C) 0.72×10^{-1} D) 0.72×10^{-2}

52) Which of the below represents the scientific notation of

0.00043

A) 0.43×10^{-1} B) 4.3×10^0

C) 4.3×10^{-4} D) 4.3×10^1

53) Which of the below represents the scientific notation of

0.0000032

A) 3.2×10^{-3} B) 0.32×10^{-3}

C) 3.2×10^{-6} D) 3.2×10^6

54) Which of the below represents the scientific notation of

0.0002

A) 2×10^{-4} B) 2×10^{-5}

C) 20×10^{-4} D) 200×10^{-4}

55) Which of the below represents the scientific notation of

0.561

A) 5.61×10^{-2} B) 5.61×10^{-3}

C) 5.61×10^{-1} D) 5.61×10^0

56) Which of the below represents the scientific notation of

0.3

A) 3×10^1 B) 3×10^2

C) 30×10^{-1} D) 3×10^{-1}

57) Which of the below represents the scientific notation of

0.018

A) 1.8×10^{-2} B) 1.8×10^0

C) 1.8×10^1 D) 0.18×10^0

58) Which of the below represents the scientific notation of

0.0049

A) 4.9×10^{-2} B) 49×10^{-3}

C) 49×10^{-4} D) 4.9×10^{-3}

59) Which of the below represents the scientific notation of

48000

A) 4.8×10^{-2} B) 4.8×10^4

C) 0.48×10^4 D) 48×10^{-2}

60) Which of the below represents the scientific notation of

0.00019

A) 1.9×10^{-4} B) 1.9×10^2

C) 1.9×10^4 D) 1.9×10^1

Grade 8

Vol 1 Week 6 Scientific Notation

61) Which of the below represents the scientific notation of

60000

A) 6×10^{-1} B) 6×10^4

C) 6×10^3 D) 6×10^2

62) Which of the below represents the scientific notation of

5200000000

A) 5.2×10^{-9} B) 5.2×10^9

C) 5.2×10^8 D) 5.2×10^{-8}

63) Which of the below represents the scientific notation of

0.0000000053

A) 5.3×10^7 B) 5.3×10^{-9}

C) 5.3×10^9 D) 5.3×10^{-8}

64) Which of the below represents the scientific notation of

13000

A) 1.3×10^{-1} B) 1.3×10^{-2}

C) 1.3×10^{-3} D) 1.3×10^4

65) Which of the below represents the scientific notation of

51.6

A) 5.16×10^{-2} B) 5.16×10^0

C) 5.16×10^1 D) 5.16×10^{-1}

66) Which of the below represents the scientific notation of

0.00000005

A) 5×10^{-8} B) 5×10^8

C) 0.5×10^8 D) 0.5×10^6

67) Which of the below represents the scientific notation of

0.0000000092

A) 9.2×10^{-9} B) 9.2×10^{10}

C) 9.2×10^9 D) 9.2×10^{11}

68) Which of the below represents the scientific notation of

0.0000002

A) 0.2×10^{-7} B) 0.2×10^{-5}

C) 0.2×10^{-6} D) 2×10^{-7}

69) Which of the below represents the scientific notation of

48200000

A) 4.82×10^7 B) 4.82×10^6

C) 4.82×10^{-7} D) 0.482×10^6

70) Which of the below represents the scientific notation of

44000000000

A) 0.044×10^7 B) 4.4×10^7

C) 4.4×10^{10} D) 0.44×10^7

Grade 8

Vol 1 Week 6 Scientific Notation

71) Which of the below represents the scientific notation of

0.0000000004

A) 4×10^{-10} B) 4×10^9

C) 4×10^8 D) 4×10^{-9}

72) Which of the below represents the scientific notation of

0.66

A) 66×10^{-1} B) 66×10^0

C) 6.6×10^{-1} D) 6.6×10^0

73) Which of the below represents the scientific notation of

900000

A) 9×10^5 B) 9×10^{-6}

C) 9×10^6 D) 9×10^{-4}

74) Which of the below represents the scientific notation of

0.000059

A) 0.59×10^{-5} B) 5.9×10^{-4}

C) 5.9×10^{-5} D) 0.59×10^4

75) Which of the below represents the scientific notation of

0.6

A) 6×10^3 B) 6×10^{-2}

C) 6×10^{-3} D) 6×10^{-1}

76) Which of the below represents the scientific notation of

0.0000012

A) 1.2×10^7 B) 1.2×10^6

C) 1.2×10^{-6} D) 1.2×10^5

77) Which of the below represents the scientific notation of

600

A) 6×10^{-2} B) 0.6×10^{-1}

C) 6×10^2 D) 6×10^{-1}

78) Which of the below represents the scientific notation of

0.0009

A) 9×10^{-4} B) 0.09×10^1

C) 0.09×10^{-4} D) 0.9×10^{-4}

79) Which of the below represents the scientific notation of

56.1

A) 5.61×10^2 B) 5.61×10^1

C) 5.61×10^{-2} D) 5.61×10^{-1}

80) Which of the below represents the scientific notation of

27000000

A) 2.7×10^7 B) 0.27×10^7

C) 2.7×10^1 D) 27×10^7

Grade 8

**Vol 1
Week 6
Scientific Notation**

81) Which of the below represents the scientific notation of

0.0000000414

A) 0.414×10^{-9} B) 4.14×10^{5}

C) 4.14×10^{-9} D) 4.14×10^{-8}

82) Which of the below represents the scientific notation of

0.9

A) 90×10^{2} B) 9×10^{-1}

C) 900×10^{2} D) 90×10^{-1}

83) Which of the below represents the scientific notation of

321

A) 3.21×10^{1} B) 3.21×10^{2}

C) 3.21×10^{0} D) 3.21×10^{-1}

84) Which of the below represents the scientific notation of

0.000894

A) 8.94×10^{-4} B) 8.94×10^{4}

C) 89.4×10^{4} D) 8.94×10^{-5}

85) Which of the below represents the scientific notation of

0.00000002

A) 2×10^{-8} B) 20×10^{8}

C) 20×10^{9} D) 2×10^{8}

86) Which of the below represents the scientific notation of

0.000000004

A) 4×10^{4} B) 4×10^{3}

C) 4×10^{-8} D) 4×10^{-9}

87) Which of the below represents the scientific notation of

2.3

A) 2.3×10^{1} B) 0.23×10^{0}

C) 0.23×10^{-1} D) 2.3×10^{0}

88) Which of the below represents the scientific notation of

8

A) 0.08×10^{-1} B) 0.8×10^{0}

C) 0.8×10^{-1} D) 8×10^{0}

89) Which of the below represents the scientific notation of

0.0000037

A) 37×10^{-5} B) 3.7×10^{-6}

C) 3.7×10^{-5} D) 37×10^{-6}

90) Which of the below represents the scientific notation of

0.0008

A) 8×10^{-3} B) 8×10^{4}

C) 8×10^{-4} D) 0.8×10^{-3}

©All rights reserved-Math-Knots LLC., VA-USA   www.math-knots.com | www.a4ace.com

Grade 8

Vol 1 Week 6 Scientific Notation

91) Which of the below represents the scientific notation of

150

A) 1.5×10^{-2} B) 0.15×10^{-2}

C) 0.15×10^{0} D) 1.5×10^{2}

92) Which of the below represents the scientific notation of

41

A) 4.1×10^{-2} B) 4.1×10^{2}

C) 4.1×10^{1} D) 41×10^{2}

93) Which of the below represents the scientific notation of

0.00005

A) 5×10^{-4} B) 5×10^{-5}

C) 5×10^{-2} D) 5×10^{4}

94) Which of the below represents the scientific notation of

39000000

A) 390×10^{-7} B) 390×10^{7}

C) 39×10^{7} D) 3.9×10^{7}

95) Which of the below represents the scientific notation of

0.0098

A) 9.8×10^{3} B) 9.8×10^{-3}

C) 98×10^{-2} D) 98×10^{-3}

96) Which of the below represents the scientific notation of

3100000

A) 3.1×10^{6} B) 31×10^{7}

C) 31×10^{-6} D) 31×10^{6}

97) Which of the below represents the scientific notation of

46000000

A) 46×10^{7} B) 4.6×10^{7}

C) 0.46×10^{7} D) 46×10^{8}

98) Which of the below represents the scientific notation of

20000

A) 2×10^{4} B) 20×10^{4}

C) 2000×10^{-2} D) 200×10^{4}

99) Which of the below represents the scientific notation of

0.000000023

A) 2.3×10^{-8} B) 23×10^{7}

C) 23×10^{-7} D) 2.3×10^{-7}

100) Which of the below represents the scientific notation of

0.000275

A) 0.275×10^{4} B) 0.275×10^{-4}

C) 27.5×10^{-4} D) 2.75×10^{-4}

Grade 8

Vol 1 Week 6 Scientific Notation

101) Simplify and express in scientific notation.

$$(9 \times 10^1)(8.59 \times 10^2)$$

A) 0.7731×10^4 B) 0.7731×10^5

C) 0.07731×10^5 D) 7.731×10^4

102) Simplify and express in scientific notation.

$$(5 \times 10^1)(3.6 \times 10^3)$$

A) 18×10^5 B) 1.389×10^{-2}

C) 1.389×10^2 D) 1.8×10^5

103) Simplify and express in scientific notation.

$$(6.31 \times 10^0)(2.33 \times 10^{-6})$$

A) 1.47×10^{-5} B) 14.7×10^{-5}

C) 14.7×10^{-3} D) 14.7×10^{-4}

104) Simplify and express in scientific notation.

$$(4.74 \times 10^3)(3.1 \times 10^0)$$

A) 146.9×10^4 B) 14.69×10^4

C) 1.469×10^4 D) 146.9×10^{-4}

105) Simplify and express in scientific notation.

$$(5.2 \times 10^5)(8.7 \times 10^4)$$

A) 4.524×10^{10} B) 4.524×10^{-9}

C) 4.524×10^{-10} D) 4.524×10^3

106) Simplify and express in scientific notation.

$$(3 \times 10^5)(2.56 \times 10^1)$$

A) 7.68×10^6 B) 7.68×10^{-2}

C) 7.68×10^3 D) 7.68×10^{-3}

107) Simplify and express in scientific notation.

$$(5.4 \times 10^{-6})(3 \times 10^0)$$

A) 1.62×10^{-5} B) 1.8×10^6

C) 1.8×10^{-6} D) 1.62×10^{-4}

108) Simplify and express in scientific notation.

$$(7.9 \times 10^5)(8.8 \times 10^1)$$

A) 69.52×10^6 B) 6.952×10^6

C) 6.952×10^{-6} D) 6.952×10^7

109) Simplify and express in scientific notation.

$$(2.4 \times 10^2)(9.6 \times 10^{-5})$$

A) 2.5×10^6 B) 0.2304×10^{-2}

C) 25×10^6 D) 2.304×10^{-2}

110) Simplify and express in scientific notation.

$$(2.46 \times 10^{-1})(1.34 \times 10^{-1})$$

A) 3.296×10^2 B) 3.296×10^{-3}

C) 1.836×10^0 D) 3.296×10^{-2}

Grade 8

Vol 1 Week 6 Scientific Notation

111) Simplify and express in scientific notation.

$$(1.1 \times 10^4)(1.7 \times 10^{-4})$$

A) 1.87×10^0 B) 1.87×10^1
C) 6.471×10^{-7} D) 6.471×10^7

112) Simplify and express in scientific notation.

$$(2 \times 10^{-2})(6.5 \times 10^{-1})$$

A) 1.3×10^{-1} B) 1.3×10^1
C) 1.3×10^{-2} D) 3.077×10^{-2}

113) Simplify and express in scientific notation.

$$(6.52 \times 10^4)(5 \times 10^0)$$

A) 3.26×10^6 B) 326×10^5
C) 3.26×10^5 D) 32.6×10^5

114) Simplify and express in scientific notation.

$$(5.7 \times 10^6)(9 \times 10^4)$$

A) 5.13×10^{-10} B) 5.13×10^{11}
C) 51.3×10^{-10} D) 5.13×10^{10}

115) Simplify and express in scientific notation.

$$(2 \times 10^2)(3.3 \times 10^2)$$

A) 6.061×10^{-1} B) 6.6×10^4
C) 66×10^4 D) 6.061×10^0

116) Simplify and express in scientific notation.

$$(8.1 \times 10^{-2})(3 \times 10^{-3})$$

A) 2.43×10^2 B) 2.43×10^{-1}
C) 2.43×10^{-4} D) 2.43×10^{-3}

117) Simplify and express in scientific notation.

$$(3.3 \times 10^1)(4.5 \times 10^2)$$

A) 1.485×10^4 B) 1.485×10^{-4}
C) 0.1485×10^4 D) 0.01485×10^4

118) Simplify and express in scientific notation.

$$(7 \times 10^2)(5.96 \times 10^{-3})$$

A) 41.72×10^0 B) 4.172×10^0
C) 1.174×10^5 D) 1.174×10^{-5}

119) Simplify and express in scientific notation.

$$(5.5 \times 10^{-4})(5 \times 10^6)$$

A) 1.1×10^{-10} B) 11×10^{-10}
C) 2.75×10^3 D) 1.1×10^{10}

120) Simplify and express in scientific notation.

$$(4 \times 10^5)(9.5 \times 10^3)$$

A) 3.8×10^9 B) 3.8×10^3
C) 4.211×10^1 D) 3.8×10^8

©All rights reserved-Math-Knots LLC., VA-USA   www.math-knots.com | www.a4ace.com

Grade 8

Vol 1 Week 6 Scientific Notation

121) Simplify and express in scientific notation.

$$(4.8 \times 10^{-3})(1.1 \times 10^4)$$

A) 4.364×10^{-7} B) 5.28×10^{-1}

C) 0.528×10^{-1} D) 5.28×10^1

122) Simplify and express in scientific notation.

$$(2.7 \times 10^{-3})(9 \times 10^{-1})$$

A) 2.43×10^4 B) 2.43×10^{-4}

C) 2.43×10^{-3} D) 24.3×10^{-3}

123) Simplify and express in scientific notation.

$$(7.5 \times 10^{-6})(7.9 \times 10^{-5})$$

A) 5.925×10^{-10} B) 5.925×10^{-9}

C) 9.494×10^{-2} D) 9.494×10^{-1}

124) Simplify and express in scientific notation.

$$(2.54 \times 10^{-5})(8 \times 10^1)$$

A) 2.032×10^{-3} B) 2.032×10^{-4}

C) 3.175×10^{-7} D) 0.2032×10^{-4}

125) Simplify and express in scientific notation.

$$(9.97 \times 10^5)(6.72 \times 10^5)$$

A) 6.7×10^{-11} B) 6.7×10^{10}

C) 6.7×10^{11} D) 1.484×10^0

126) Simplify and express in scientific notation.

$$(8.1 \times 10^{-4})(3.3 \times 10^{-1})$$

A) 2.673×10^3 B) 2.673×10^{-4}

C) 2.673×10^2 D) 26.73×10^3

127) Simplify and express in scientific notation.

$$(1.93 \times 10^{-1})(6 \times 10^6)$$

A) 3.217×10^{-8} B) 1.158×10^6

C) 11.58×10^6 D) 11.58×10^0

128) Simplify and express in scientific notation.

$$(2 \times 10^4)(4.8 \times 10^{-1})$$

A) 0.96×10^{-2} B) 0.96×10^{-3}

C) 9.6×10^3 D) 0.96×10^3

129) Simplify and express in scientific notation.

$$(6.6 \times 10^1)(5.26 \times 10^{-2})$$

A) 0.3472×10^2 B) 3.472×10^1

C) 3.472×10^0 D) 3.472×10^2

130) Simplify and express in scientific notation.

$$(5.5 \times 10^6)(3.6 \times 10^{-2})$$

A) 1.528×10^{10} B) 1.98×10^5

C) 1.528×10^8 D) 1.528×10^9

Grade 8

Vol 1 Week 6 Scientific Notation

131) Simplify and express in scientific notation.

$$(3 \times 10^{-4})(1.7 \times 10^{-3})$$

A) 5.1×10^7 B) 5.1×10^{-7}

C) 1.765×10^{-2} D) 1.765×10^{-1}

132) Simplify and express in scientific notation.

$$(5.9 \times 10^{-4})(2.1 \times 10^6)$$

A) 12.39×10^3 B) 12.39×10^{-3}

C) 12.39×10^{-2} D) 1.239×10^3

133) Simplify and express in scientific notation.

$$(9.6 \times 10^4)(2 \times 10^{-2})$$

A) 1.92×10^{-1} B) 1.92×10^4

C) 1.92×10^{-2} D) 1.92×10^3

134) Simplify and express in scientific notation.

$$(6.7 \times 10^5)(4.56 \times 10^0)$$

A) 0.3055×10^7 B) 0.3055×10^8

C) 3.055×10^6 D) 3.055×10^7

135) Simplify and express in scientific notation.

$$(8.7 \times 10^{-3})(3 \times 10^0)$$

A) 2.61×10^{-2} B) 2.9×10^{-3}

C) 0.29×10^{-3} D) 2.61×10^{-3}

136) Simplify and express in scientific notation.

$$(8.98 \times 10^{-5})(7 \times 10^6)$$

A) 62.86×10^1 B) 6.286×10^{-2}

C) 6.286×10^1 D) 6.286×10^2

137) Simplify and express in scientific notation.

$$(1.8 \times 10^0)(9.7 \times 10^4)$$

A) 1.746×10^1 B) 1.856×10^5

C) 1.856×10^{-5} D) 1.746×10^5

138) Simplify and express in scientific notation.

$$(7.35 \times 10^1)(8 \times 10^3)$$

A) 58.8×10^3 B) 5.88×10^5

C) 58.8×10^4 D) 5.88×10^4

139) Simplify and express in scientific notation.

$$(1.6 \times 10^{-3})(5 \times 10^3)$$

A) 8×10^0 B) 3.2×10^{-7}

C) 32×10^{-7} D) 8×10^1

140) Simplify and express in scientific notation.

$$(4.2 \times 10^0)(6.8 \times 10^{-1})$$

A) 6.176×10^0 B) 2.856×10^0

C) 61.76×10^0 D) 2.856×10^1

Grade 8

Vol 1
Week 7
Scientific Notation

1) Simplify and express in scientific notation.

$$(7 \times 10^0)(3.42 \times 10^6)$$

A) 2.394×10^8 B) 2.394×10^9

C) 2.047×10^{-6} D) 2.394×10^7

2) Simplify and express in scientific notation.

$$(7 \times 10^6)(3.74 \times 10^6)$$

A) 2.618×10^{12} B) 2.618×10^{-13}

C) 2.618×10^{13} D) 2.618×10^{-12}

3) Simplify and express in scientific notation.

$$(7.6 \times 10^2)(5.9 \times 10^3)$$

A) 1.288×10^{-1} B) 4.484×10^6

C) 1.288×10^0 D) 4.484×10^5

4) Simplify and express in scientific notation.

$$(4.1 \times 10^1)(6.1 \times 10^6)$$

A) 6.721×10^{-6} B) 67.21×10^{-6}

C) 2.501×10^9 D) 2.501×10^8

5) Simplify and express in scientific notation.

$$(5.4 \times 10^{-6})(6 \times 10^{-1})$$

A) 0.324×10^{-6} B) 9×10^{-6}

C) 0.324×10^{-5} D) 3.24×10^{-6}

6) Simplify and express in scientific notation.

$$\frac{2.7 \times 10^5}{8 \times 10^{-1}}$$

A) 2.16×10^5 B) 3.375×10^6

C) 2.16×10^{-5} D) 3.375×10^5

7) Simplify and express in scientific notation.

$$(8.69 \times 10^{-1})(2.3 \times 10^2)$$

A) 1.999×10^1 B) 0.1999×10^0

C) 1.999×10^2 D) 0.1999×10^1

8) Simplify and express in scientific notation.

$$(1.6 \times 10^{-4})(7.69 \times 10^4)$$

A) 12.3×10^1 B) 2.081×10^9

C) 2.081×10^{-9} D) 1.23×10^1

9) Simplify and express in scientific notation.

$$(1.29 \times 10^{-2})(3 \times 10^{-4})$$

A) 4.3×10^1 B) 38.7×10^6

C) 38.7×10^{-6} D) 3.87×10^{-6}

10) Simplify and express in scientific notation.

$$(7 \times 10^{-1})(7.29 \times 10^{-2})$$

A) 51.03×10^{-4} B) 5.103×10^{-3}

C) 5.103×10^{-4} D) 5.103×10^{-2}

Grade 8

Vol 1 Week 7 Scientific Notation

11) Simplify and express in scientific notation.

$$(1.2 \times 10^2)(6.9 \times 10^1)$$

A) 1.739×10^0 B) 8.28×10^3

C) 1.739×10^1 D) 0.828×10^3

12) Simplify and express in scientific notation.

$$\frac{4 \times 10^2}{2.67 \times 10^4}$$

A) 1.498×10^{-2} B) 1.068×10^7

C) 106.8×10^7 D) 10.68×10^7

13) Simplify and express in scientific notation.

$$\frac{7.02 \times 10^4}{1.63 \times 10^{-1}}$$

A) 4.307×10^2 B) 4.307×10^5

C) 4.307×10^4 D) 4.307×10^{-2}

14) Simplify and express in scientific notation.

$$\frac{9.3 \times 10^{-2}}{9.9 \times 10^5}$$

A) 0.9207×10^4 B) 9.394×10^{-9}

C) 9.394×10^{-8} D) 9.207×10^4

15) Simplify and express in scientific notation.

$$\frac{5.6 \times 10^6}{9.26 \times 10^0}$$

A) 5.186×10^{-7} B) 60.48×10^5

C) 6.048×10^5 D) 5.186×10^7

16) Simplify and express in scientific notation.

$$\frac{7 \times 10^1}{7 \times 10^4}$$

A) 10^{-3} B) 4.9×10^7

C) 4.9×10^6 D) 10^3

17) Simplify and express in scientific notation.

$$\frac{2.5 \times 10^{-4}}{6.9 \times 10^{-5}}$$

A) 1.725×10^{-7} B) 1.725×10^{-8}

C) 3.623×10^{-1} D) 3.623×10^0

18) Simplify and express in scientific notation.

$$\frac{2.4 \times 10^{-6}}{3 \times 10^{-2}}$$

A) 8×10^6 B) 8×10^5

C) 7.2×10^{-8} D) 8×10^{-5}

19) Simplify and express in scientific notation.

$$\frac{6.1 \times 10^1}{6 \times 10^{-1}}$$

A) 1.017×10^{-2} B) 1.017×10

C) 1.017×10^2 D) 1.017×10^0

20) Simplify and express in scientific notation.

$$\frac{4.51 \times 10^{-3}}{7 \times 10^5}$$

A) 3.157×10^{-3} B) 6.443×10^{-10}

C) 3.157×10^3 D) 6.443×10^{-9}

Grade 8

Vol 1 Week 7 Scientific Notation

21) Simplify and express in scientific notation.
$$\frac{7.2 \times 10^3}{9.58 \times 10^{-3}}$$

A) 7.516×10^5 B) 0.7516×10^4

C) 7.516×10^4 D) 6.898×10^1

22) Simplify and express in scientific notation.
$$\frac{8.6 \times 10^3}{6.7 \times 10^5}$$

A) 1.284×10^{-1} B) 0.1284×10^{-1}

C) 1.284×10^{-2} D) 5.762×10^9

23) Simplify and express in scientific notation.
$$\frac{9.01 \times 10^{-1}}{7 \times 10^1}$$

A) 6.307×10^0 B) 1.287×10^2

C) 1.287×10^{-2} D) 6.307×10^1

24) Simplify and express in scientific notation.
$$\frac{3.4 \times 10^{-3}}{9.8 \times 10^{-6}}$$

A) 3.469×10^2 B) 0.3469×10^{-1}

C) 0.3469×10^{-2} D) 0.3469×10^2

25) Simplify and express in scientific notation.
$$\frac{1.9 \times 10^0}{9.98 \times 10^3}$$

A) 0.1904×10^5 B) 1.904×10^{-4}

C) 1.904×10^5 D) 1.904×10^{-5}

26) Simplify and express in scientific notation.
$$\frac{4.63 \times 10^4}{2.3 \times 10^2}$$

A) 2.013×10^2 B) 20.13×10^{-2}

C) 1.065×10^7 D) 20.13×10^2

27) Simplify and express in scientific notation.
$$\frac{9.9 \times 10^6}{1.21 \times 10^5}$$

A) 8.182×10^1 B) 0.8182×10^1

C) 1.198×10^{-12} D) 1.198×10^{12}

28) Simplify and express in scientific notation.
$$\frac{1.4 \times 10^0}{5.59 \times 10^{-6}}$$

A) 2.504×10^4 B) 2.504×10^3

C) 2.504×10^6 D) 2.504×10^5

29) Simplify and express in scientific notation.
$$\frac{5.19 \times 10^{-5}}{5.4 \times 10^6}$$

A) 2.803×10^{-2} B) 2.803×10^2

C) 9.611×10^{12} D) 9.611×10^{-12}

30) Simplify and express in scientific notation.
$$\frac{5.8 \times 10^{-1}}{1.83 \times 10^{-2}}$$

A) 3.169×10^1 B) 0.3169×10^{-1}

C) 3.169×10^{-1} D) 0.3169×10^0

Grade 8

Vol 1 Week 7 Scientific Notation

31) Simplify and express in scientific notation.
$$\frac{1.46 \times 10^{-5}}{7 \times 10^3}$$
A) 2.086×10^{-4} B) 1.022×10^{-1}
C) 1.022×10^{1} D) 2.086×10^{-9}

32) Simplify and express in scientific notation.
$$\frac{5.4 \times 10^{-3}}{4 \times 10^{-6}}$$
A) 0.0135×10^{-3} B) 0.135×10^{-3}
C) 1.35×10^{3} D) 0.135×10^{3}

33) Simplify and express in scientific notation.
$$\frac{7.42 \times 10^{-3}}{4.42 \times 10^{2}}$$
A) 0.1679×10^{-4} B) 0.1679×10^{-5}
C) 1.679×10^{-5} D) 0.1679×10^{-3}

34) Simplify and express in scientific notation.
$$\frac{9.9 \times 10^{6}}{3.8 \times 10^{-6}}$$
A) 2.605×10^{12} B) 0.2605×10^{12}
C) 0.02605×10^{12} D) 2.605×10^{11}

35) Simplify and express in scientific notation.
$$\frac{8.11 \times 10^{2}}{7.23 \times 10^{5}}$$
A) 11.22×10^{-3} B) 5.864×10^{8}
C) 1.122×10^{3} D) 1.122×10^{-3}

36) Simplify and express in scientific notation.
$$\frac{9 \times 10^{-4}}{5 \times 10^{-6}}$$
A) 1.8×10^{0} B) 0.18×10^{1}
C) 1.8×10^{2} D) 1.8×10^{1}

37) Simplify and express in scientific notation.
$$\frac{9.57 \times 10^{-3}}{5.14 \times 10^{-2}}$$
A) 0.1862×10^{-1} B) 1.862×10^{-1}
C) 0.01862×10^{0} D) 0.01862×10^{-1}

38) Simplify and express in scientific notation.
$$\frac{4.25 \times 10^{2}}{1.76 \times 10^{3}}$$
A) 2.415×10^{-1} B) 7.48×10^{5}
C) 0.2415×10^{-2} D) 0.2415×10^{-1}

39) Simplify and express in scientific notation.
$$\frac{6 \times 10^{-5}}{7.9 \times 10^{0}}$$
A) 7.595×10^{-7} B) 4.74×10^{-4}
C) 4.74×10^{4} D) 7.595×10^{-6}

40) Simplify and express in scientific notation.
$$\frac{4.3 \times 10^{5}}{4 \times 10^{1}}$$
A) 1.075×10^{4} B) 1.72×10^{7}
C) 10.75×10^{4} D) 1.72×10^{8}

©All rights reserved-Math-Knots LLC., VA-USA

Grade 8

Vol 1 Week 7 Scientific Notation

41) Simplify and express in scientific notation.
$$\frac{6.94 \times 10^{-1}}{2 \times 10^{-1}}$$
A) 3.47×10^1 B) 3.47×10^0
C) 34.7×10^0 D) 1.388×10^{-1}

42) Simplify and express in scientific notation.
$$\frac{7 \times 10^{-6}}{7 \times 10^{-3}}$$
A) 4.9×10^{-8} B) 10^3
C) 10×10^3 D) 10^{-3}

43) Simplify and express in scientific notation.
$$\frac{8.5 \times 10^{-5}}{3.01 \times 10^{-6}}$$
A) 2.824×10^{-1} B) 2.824×10^1
C) 2.558×10^{-10} D) 0.2824×10^{-1}

44) Simplify and express in scientific notation.
$$\frac{6.86 \times 10^2}{5.8 \times 10^1}$$
A) 1.183×10^2 B) 3.979×10^4
C) 1.183×10^1 D) 3.979×10^{-4}

45) Simplify and express in scientific notation.
$$\frac{9 \times 10^{-3}}{1.6 \times 10^3}$$
A) 0.5625×10^6 B) 5.625×10^6
C) 5.625×10^{-6} D) 1.44×10^1

46) Simplify and express in scientific notation.
$$\frac{6.74 \times 10^{-1}}{7.21 \times 10^2}$$
A) 4.86×10^2 B) 9.348×10^{-3}
C) 9.348×10^{-4} D) 0.486×10^2

47) Simplify and express in scientific notation.
$$\frac{5.76 \times 10^{-6}}{6.4 \times 10^5}$$
A) 9×10^{-12} B) 9×10^1
C) 9×10^{-1} D) 3.686×10^0

48) Simplify and express in scientific notation.
$$\frac{9.2 \times 10^{-2}}{3 \times 10^4}$$
A) 3.067×10^{-5} B) 2.76×10^3
C) 3.067×10^{-6} D) 3.067×10^6

49) Simplify and express in scientific notation.
$$\frac{8.5 \times 10^{-1}}{3 \times 10^6}$$
A) 0.255×10^6 B) 2.833×10^{-7}
C) 2.55×10^6 D) 2.833×10^{-8}

50) Simplify and express in scientific notation.
$$\frac{6 \times 10^{-3}}{4.87 \times 10^6}$$
A) 1.232×10^{-8} B) 2.922×10^{-4}
C) 1.232×10^{-9} D) 2.922×10^4

Vol 1
Week 7
Scientific Notation

Grade 8

51) Simplify and express in scientific notation.
$$\frac{3.32 \times 10^{-5}}{4 \times 10^5}$$
A) 8.3×10^{-11} B) 1.328×10^{-1}
C) 8.3×10^{11} D) 1.328×10^{1}

52) Simplify and express in scientific notation.
$$\frac{1.8 \times 10^{-2}}{4.9 \times 10^4}$$
A) 0.882×10^2 B) 3.673×10^{-7}
C) 3.673×10^7 D) 8.82×10^2

53) Simplify and express in scientific notation.
$$\frac{7.25 \times 10^{-5}}{2.7 \times 10^{-1}}$$
A) 1.958×10^5 B) 1.958×10^{-5}
C) 2.685×10^{-4} D) 2.685×10^{-3}

54) Simplify and express in scientific notation.
$$\frac{9.31 \times 10^{-2}}{5.5 \times 10^{-2}}$$
A) 5.121×10^{-3} B) 1.693×10^{-1}
C) 5.121×10^{-4} D) 1.693×10^{0}

55) Simplify and express in scientific notation.
$$\frac{6.6 \times 10^3}{4 \times 10^{-5}}$$
A) 1.65×10^8 B) 2.64×10^{-1}
C) 1.65×10^7 D) 1.65×10^{-7}

56) Simplify and express in scientific notation.
$$\frac{5 \times 10^0}{9 \times 10^6}$$
A) 4.5×10^7 B) 5.556×10^{-7}
C) 4.5×10^8 D) 5.556×10^5

57) Simplify and express in scientific notation.
$$(9 \times 10^1)^{-5}$$
A) 1.694×10^{-10} B) 1.694×10^{-11}
C) 16.94×10^{-10} D) 16.94×10^{-11}

58) Simplify and express in scientific notation.
$$\frac{6.02 \times 10^1}{8 \times 10^{-3}}$$
A) 75.25×10^3 B) 752.5×10^{-3}
C) 7.525×10^3 D) 752.5×10^3

59) Simplify and express in scientific notation.
$$\frac{4.57 \times 10^0}{6.3 \times 10^{-2}}$$
A) 7.254×10^1 B) 0.2879×10^{-1}
C) 0.7254×10^1 D) 2.879×10^{-1}

60) Simplify and express in scientific notation.
$$\frac{3.61 \times 10^{-1}}{1.9 \times 10^{-3}}$$
A) 1.9×10^2 B) 6.859×10^{-4}
C) 6.859×10^{-3} D) 0.19×10^2

Grade 8

Vol 1 Week 7 Scientific Notation

61) Simplify and express in scientific notation.

$$\frac{3.2 \times 10^{-3}}{3 \times 10^0}$$

A) 9.6×10^{-3} B) 1.067×10^{-3}

C) 96×10^{-3} D) 10.67×10^{-3}

62) Simplify and express in scientific notation.

$$(8.9 \times 10^{-4})^{-4}$$

A) 1.594×10^{12} B) 8.9×10^{-10}

C) 1.594×10^{-10} D) 0.1594×10^{-10}

63) Simplify and express in scientific notation.

$$(8.9 \times 10^4)^{-2}$$

A) 1.563×10^{12} B) 1.563×10^{-10}

C) 1.563×10^{10} D) 1.563×10^{11}

64) Simplify and express in scientific notation.

$$(8 \times 10^{-3})^{-4}$$

A) 8×10^{-4} B) 2.441×10^{-4}

C) 2.441×10^{-10} D) 2.441×10^{-10}

65) Simplify and express in scientific notation.

$$(5.9 \times 10^2)^4$$

A) 1.212×10^{10} B) 1.212×10^{-10}

C) 12.12×10^{10} D) 5.9×10^{11}

66) Simplify and express in scientific notation.

$$(86.68 \times 10^{-2})^{-2}$$

A) 22.41×10^2 B) 2.241×10^2

C) 22.41×10^2 D) 22.41×10^{-1}

67) Simplify and express in scientific notation.

$$(8.77 \times 10^2)^{-2}$$

A) 1.3×10^{-7} B) 1.3×10^6

C) 1.3×10^7 D) 1.3×10^{-6}

68) Simplify and express in scientific notation.

$$(7.8 \times 10^3)^2$$

A) 7.8×10^7 B) 0.78×10^7

C) 6.084×10^7 D) 6.084×10^{-7}

69) Simplify and express in scientific notation.

$$(8.9 \times 10^{-4})^4$$

A) 1.594×10^{12} B) 8.9×10^{-10}

C) 1.594×10^{-10} D) 0.1594×10^{-10}

70) Simplify and express in scientific notation.

$$(8 \times 10^1)$$

A) 4.096×10^7 B) 8×10^{-6}

C) 8×10^6 D) 4.096×10^6

71) In last ten years, approximately 7,570,000 people visited the Eifel Tower. Express this number in scientific notation?

A. 0.757×10^7
B. 757×10
C. 7.57×10^6
D. 75.7×10^5

72) If a number expressed in scientific notation is $N \times 10^3$, how large is the number?

A. Between 1,000 (included) and 10,000
B. Between 10,000 (included) and 100,000
C. Between 100,000 (included) and 1,000,000
D. Between 1,000,000 (included) and 10,000,000

73) If a number expressed in scientific notation is $N \times 10^7$, how large is the number?

A. Between 1,000 (included) and 10,000
B. Between 10,000 (included) and 100,000
C. Between 100,000 (included) and 1,000,000
D. Between 10,000,000 (included) and 100,000,000

74) If a number expressed in scientific notation is $N \times 10^{10}$, how large is the number?

A. Between 1,000 (included) and 10,000
B. Between 10,000 (included) and 100,000
C. Between 100,000 (included) and 1,000,000
D. Between 10,000,000,000 (included) and 100,000,000,000

75) Which of the following numbers has the greatest value?

A. 11.13×10^3
B. 11.135×10^2
C. 11.135×10^3
D. 111.35×10^1

76) Which of the following numbers has the least value?

A. -4.23×10^2
B. 4.50×10^{-3}
C. -4.23×10^5
D. 4.23×10^4

77) The average distance from Saturn to the Neptune is 3,076,400,000 miles. Express this number in scientific notation.

A. 3076×10^8
B. 307.6×10^9
C. 30.764×10^8
D. 3.0764×10^{11}

Grade 8

Vol 1 Week 7 Scientific Notation

78) The average distance from Venus to the Mars is 119,700,000 miles. Express this number in scientific notation.

  A. 11.97×10^7 B. 1197×10^5

  C. 1194×10^8 D. 1.197×10^5

79) The average distance from Mercury to the Venus is 50,290,000 miles. Express this number in scientific notation.

  E. 5029×10^8 F. 5029×10^5

  G. 5029×10^5 H. 50.29×10^6

80) The approximate population of India is 1.38×10^9 people. Express this number in standard notation.

  A. 138,000   B. 1,380,000,000

  C. 13,800,000   D. 138,000,000

81) The approximate population of United States is 3.31×10^{-8} people. Express this number in standard notation.

  A. 331,000,000   B. 3,310,000

  C. 33,100,000   D. 331,000

82) The approximate population of France is 65.27×10^6 people. Express this number in standard notation.

  A. 652,700,000   B. 652,700

  C. 6,527,000   D. 65,270,000

83) The approximate population of Malaysia is 3.237×10^7 people. Express this number in standard notation.

  A. 237,000   B. 323,700,000

  C. 32,370,000   D. 3,237,000

84) The approximate population of Iceland is 0.000341×10^9 people. Express this number in standard notation.

  A. 341,000   B. 3,410,000

  C. 34,100,000   D. 341,000,000

85) Mercury is approximately 50.3×10^{-6} kilometers from the Venus. The speed of light is approximately 3×10^5 kilometers per second. Divide the distance by the speed of light to determine the approximate number of seconds it takes light to travel from the Sun to Mercury.

  A. 2 seconds   B. 20 seconds

  C. 170 seconds   D. 1,680 seconds

Grade 8

Vol 1 Week 7 Scientific Notation

86) Earth is approximately 7.8×10^{-7} kilometers from the Mars. The speed of light is approximately 3×10^{5} kilometers per second. Divide the distance by the speed of light to determine approximate number of seconds it takes light to travel from the Sun to Mercury.

A. 6 seconds B. 30 seconds

C. 120 seconds D. 260 seconds

87) Mars is approximately 2.7×10^{-9} kilometers from the Uranus. The speed of light is approximately 3×10^{5} kilometers per second. Divide the distance by the speed of light to determine the approximate number of seconds it takes light to travel from the Sun to Mercury.

A. 9 seconds B. 900 seconds

C. 9 seconds D. 9,000 seconds

88) Chicago is approximately 9.4×10^{2} miles from houston. Dora drives 7×10^{1} miles per hour from Washington to Utah. Divide the distance by the speed to determine the approximate number of hours it takes Mary to travel from Washington to Utah

A. 12 hours B. 14 hours

C. 10 hours D. 20 hours

89) Maimi is approximately 4.9×10^{-3} miles from Hawaii. Mary drives 8×10 miles per hour from Washinton to Utah. Divide the distance by the speed to determine the approximate number of hours it takes Mary to travel from Washington to Utah.

A. 45 hours B. 50 hours

C. 74 hours D. 62 hours

90) A wood plank is approximately 1×10^{-2} meters thick. What fraction of a meter is this?

A. $\dfrac{1}{100}$ B. $\dfrac{1}{1000}$

C. $\dfrac{1}{10000}$ D. $\dfrac{1}{100000}$

91) A sheet of paper is approximately 1×10^{-6} meters thick. What fraction of a meter is this?

A. $\dfrac{1}{100}$ B. $\dfrac{1}{1000}$

C. $\dfrac{1}{10000}$ D. $\dfrac{1}{100000}$

Grade 8

Vol 1
Week 8
Assessment 1

1) The $\sqrt{637}$ lies between which two integers?

2) The $\sqrt{693}$ lies between which two integers?

3) Simplify to only positive exponents.

$$\left(\frac{(2^3)^{-3} \cdot 2^{-4}}{(2^3)^{-1}}\right)^3$$

A) 2^6 B) $\dfrac{1}{2^{30}}$

C) $\dfrac{1}{2^{24}}$ D) 1

4) Simplify the below

$$\sqrt[4]{32x^2}$$

A) $2x\sqrt[4]{4x}$ B) $2\sqrt[4]{2x^2}$

C) $2x\sqrt[4]{3x}$ D) $3x\sqrt[4]{4x}$

5) Simplify to only positive exponents.

$$\frac{7^0}{7^7}$$

A) 7^2 B) 7^{10}

C) 7^{13} D) $\dfrac{1}{7^7}$

6) Simplify to only positive exponents.

$$(2^4)^{-4}$$

A) $\dfrac{1}{2^{16}}$ B) 2^6

C) $\dfrac{1}{2^6}$ D) $\dfrac{1}{2^2}$

7) $-\sqrt{\dfrac{3600}{144}} = ?$

8) Which option has the same value as the below expression?

$$\frac{2\sqrt{4}}{\sqrt{20}}$$

A) $\dfrac{\sqrt{30}}{20}$ B) $\dfrac{2\sqrt{5}}{5}$

C) $\dfrac{\sqrt{21}}{56}$ D) $\dfrac{\sqrt{10}}{5}$

9) Which option has the same value as the below expression?

$$\frac{\sqrt{3}}{\sqrt{8}}$$

A) $\dfrac{\sqrt{15}}{35}$ B) $\dfrac{2\sqrt{6}}{3}$

C) $\dfrac{\sqrt{6}}{4}$ D) $\dfrac{18\sqrt{14}}{7}$

Grade 8 — Vol 1, Week 8, Assessment 1

10) Which option is equivalent to the below expression?

$$-3\sqrt{54} - 5\sqrt{54}$$

A) $-33\sqrt{6}$ B) $-24\sqrt{6}$

C) $-42\sqrt{6}$ D) $-57\sqrt{6}$

11) Find the value of x for the below

$$\sqrt{864}$$

A) $12\sqrt{6}$ B) $11\sqrt{15}$

C) $3\sqrt{6}$ D) $4\sqrt{15}$

12) Simplify to only positive exponents.

$$3 \cdot 3^4$$

A) $\dfrac{1}{3^6}$ B) $\dfrac{1}{3^8}$

C) 3^5 D) $\dfrac{1}{3^{16}}$

13) Simplify to only positive exponents.

$$(2x^{-30})^{10} \cdot x^{28}$$

A) x^8 B) $\dfrac{2048}{x^{275}}$

C) $128x^{320}$ D) $\dfrac{1024}{x^{272}}$

14) Which option is equivalent to the below expression?

$$\sqrt{2} \cdot \sqrt{5}$$

A) 10   B) $\sqrt{10}$

C) $\sqrt{7}$ D) $\sqrt{210}$

15) Which option is equivalent to the below expression?

$$\sqrt{70} \cdot \sqrt{21}$$

A) 1470   B) $\sqrt{105}$

C) $\sqrt{91}$ D) $7\sqrt{30}$

16) Which option is equivalent to the below expression?

$$\sqrt{15} \cdot 9\sqrt{45}$$

A) $2\sqrt{15}$ B) $15\sqrt{3}$

C) 675   D) $135\sqrt{3}$

17) $-\sqrt{\dfrac{4}{3600}} = ?$

18) Which option is equivalent to the below expression?

$$\dfrac{\sqrt{7}}{\sqrt{2}}$$

A) $\dfrac{\sqrt{14}}{2}$ B) $\dfrac{8\sqrt{3}}{9}$

C) $\dfrac{\sqrt{3}}{30}$ D) $\dfrac{\sqrt{14}}{7}$

19) Simplify to only positive exponents.

$$(v^6)^{16} \cdot v^{25}$$

A) $\dfrac{4}{v^{538}}$ B) v^{484}

C) $\dfrac{2}{v^{128}}$ D) v^{121}

Grade 8

Vol 1 Week 8 Assessment 1

20) Find the value of x for the below

$$\sqrt{392}$$

A) $11\sqrt{14}$ B) $7\sqrt{3}$

C) $14\sqrt{2}$ D) $4\sqrt{2}$

21) Simplify to only positive exponents.

$$\frac{2^4 \cdot 2^2}{\left(2^{-4}\right)^4}$$

A) 2^{22} B) $\frac{1}{2^2}$

C) 2^{16} D) 1

22) Simplify the below

$$\sqrt[3]{-16b^3}$$

A) $2b\sqrt[3]{3}$ B) $5\sqrt[3]{2b}$

C) $-2b\sqrt[3]{2}$ D) $2b\sqrt[3]{5b^2}$

23) $\sqrt{\dfrac{25}{6400}} = ?$

24) Which option is equivalent to the below expression?

$$\frac{\sqrt{5}}{3\sqrt{2}}$$

A) $\dfrac{\sqrt{10}}{6}$ B) $\dfrac{3\sqrt{10}}{5}$

C) $\dfrac{\sqrt{6}}{2}$ D) $\dfrac{3\sqrt{10}}{40}$

25) Which option has the same value as the below expression?

$$\frac{3\sqrt{40}}{\sqrt{15}}$$

A) $\dfrac{2\sqrt{5}}{5}$ B) $\dfrac{\sqrt{6}}{12}$

C) $\dfrac{9\sqrt{10}}{2}$ D) $2\sqrt{6}$

26) Simplify to only positive exponents.

$$\frac{2x^6}{15x^8 \cdot 7x^9 \cdot 4x^{12}}$$

A) $\dfrac{65x^{21}}{56}$ B) $\dfrac{1}{210x^{23}}$

C) $120x^{16}$ D) $6x^9$

Grade 8

Vol 1
Week 8
Assessment 1

27) Simplify the below

$$\sqrt[5]{96b^2}$$

A) $-2\sqrt[5]{3b^2}$ B) $2b\sqrt[5]{4b^2}$
C) $2\sqrt[5]{3b^2}$ D) $2b\sqrt[5]{3b}$

28) Simplify the below

$$\sqrt[3]{135x^2}$$

A) $-3x\sqrt[3]{2x}$ B) $5x\sqrt[3]{3}$
C) $2x\sqrt[3]{2x^2}$ D) $3\sqrt[3]{5x^2}$

29) Simplify to only positive exponents.

$$(2^3)^0$$

A) $\dfrac{1}{2^6}$ B) 1
C) $\dfrac{1}{2^4}$ D) 2^6

30) $-\sqrt{\dfrac{625}{10000}} = ?$

31) Simplify the below

$$\sqrt[3]{108m^3}$$

A) $3m\sqrt[3]{4}$ B) $2m\sqrt[3]{5m^2}$
C) $4\sqrt[3]{2m^2}$ D) $3m\sqrt[3]{4m^2}$

32) Simplify to only positive exponents.

$$\dfrac{3n^{12}}{8n^8 \cdot 4n^{-7}}$$

A) $\dfrac{4n^{22}}{7}$ B) $\dfrac{22n^3}{3}$
C) $\dfrac{3n^{11}}{32}$ D) $\dfrac{n}{8}$

33) Simplify to only positive exponents.

$$\dfrac{15x^0}{5x^{-2} \cdot 8x^{15}}$$

A) $\dfrac{1}{4x^5}$ B) $\dfrac{x^5}{6}$
C) $\dfrac{21x^{10}}{2}$ D) $\dfrac{3}{8x^{13}}$

34) Simplify to only positive exponents.

$$\dfrac{(2x^5)^{-6}}{(x^{-7})^3}$$

A) x^{84} B) $32x^{20}$
C) $\dfrac{x^{21}}{8}$ D) $\dfrac{1}{64x^9}$

35) Simplify to only positive exponents.

$$\left(\dfrac{n^{-3}}{(2n^7)^5}\right)^0$$

A) $\dfrac{1}{n^{13}}$ B) 1
C) $\dfrac{n^{12}}{2}$ D) n^{56}

Grade 8

Vol 1
Week 8
Assessment 1

36) Simplify to only positive exponents.

$$8^4 \cdot 8^4$$

A) 8^{11} B) 8^8

C) $\dfrac{1}{8^{13}}$ D) 1

37) Which option has the same value as the below expression?

$$5\sqrt{10} + 2\sqrt{90}$$

A) $11\sqrt{10}$ B) $23\sqrt{10}$

C) $17\sqrt{10}$ D) $28\sqrt{10}$

38) Simplify to only positive exponents.

$$3^{-8} \cdot 3^6$$

A) 1 B) 3^2

C) $\dfrac{1}{3^2}$ D) $\dfrac{1}{3^7}$

39) Simplify to only positive exponents.

$$2^{-2} \cdot (2^4)^3$$

A) 2^8 B) 2^{10}

C) $\dfrac{1}{2^6}$ D) 1

40) Which option is equivalent to the below expression?

$$3\sqrt{6} - \sqrt{6}$$

A) $2\sqrt{6}$ B) $11\sqrt{6}$

C) $5\sqrt{6}$ D) $8\sqrt{6}$

41) Simplify to only positive exponents.

$$2^3 \cdot 2^{-2}$$

A) $\dfrac{1}{2}$ B) 2

C) 1 D) 2^{10}

42) Simplify to only positive exponents.

$$3 \cdot 3^6$$

A) $\dfrac{1}{3^2}$ B) 3^7

C) 3^9 D) 3^3

43) Simplify to only positive exponents.

$$\dfrac{(b^{-1})^4}{b^6}$$

A) $32b^3$ B) $\dfrac{1}{b^{10}}$

C) 1 D) $8b^9$

44) Which option has the same value as the below expression?

$$4\sqrt{6} \cdot \sqrt{21}$$

A) $12\sqrt{14}$ B) $3\sqrt{14}$

C) 126 D) $3\sqrt{3}$

Grade 8

Vol 1 — Week 8 — Assessment 1

45) Simplify to only positive exponents.

$$(3^2)^{-3}$$

A) $\dfrac{1}{3^6}$ B) 3^4
C) 3^{12} D) 3^8

46) Simplify to only positive exponents.

$$(4^0)^{-2}$$

A) $\dfrac{1}{4}$ B) 4^2
C) 4^{12} D) 1

47) Simplify to only positive exponents.

$$\dfrac{3^0}{3^7}$$

A) $\dfrac{1}{3^6}$ B) 3^2
C) 3^{10} D) $\dfrac{1}{3^7}$

48) Simplify the below

$$\sqrt[3]{128n}$$

A) $4\sqrt[3]{2n}$ B) $2\sqrt[3]{3n}$
C) $3\sqrt[3]{5n}$ D) $4n\sqrt[3]{3}$

49) Simplify to only positive exponents.

$$(4^4)^{-1}$$

A) 4^8 B) 4^2
C) $\dfrac{1}{4^8}$ D) $\dfrac{1}{4^4}$

50) Simplify to only positive exponents.

$$\dfrac{9^3}{9^2}$$

A) $\dfrac{1}{9^8}$ B) 9^7
C) 9^9 D) 9

51) Simplify to only positive exponents.

$$\dfrac{6^{-10}}{6^6}$$

A) $\dfrac{1}{6^4}$ B) 6^2
C) $\dfrac{1}{6^{16}}$ D) $\dfrac{1}{6^5}$

52) Simplify to only positive exponents.

$$\dfrac{8^{-10}}{8^7}$$

A) $\dfrac{1}{8^{17}}$ B) $\dfrac{1}{8}$
C) 8^3 D) 8^{10}

Grade 8

Vol 1　Week 8　Assessment 1

53) Simplify to only positive exponents.

$$\frac{10^0}{10^5 \cdot 10^0}$$

A) $\dfrac{1}{10^6}$ B) $\dfrac{1}{10}$

C) 10^{20} D) $\dfrac{1}{10^5}$

54) Simplify to only positive exponents.

$$\frac{6^7}{6^{10} \cdot 6^{-10} \cdot 6^0}$$

A) $\dfrac{1}{6^{12}}$ B) 6^{12}

C) $\dfrac{1}{6^{22}}$ D) 6^7

55) Simplify to only positive exponents.

$$\frac{2^{-7}}{2^4 \cdot 2^{-1}}$$

A) 2^5 B) 2^{15}

C) $\dfrac{1}{2^5}$ D) $\dfrac{1}{2^{10}}$

56) Simplify to only positive exponents.

$$\left(2 \cdot 2^{-1}\right)^{-2}$$

A) $\dfrac{1}{2^3}$ B) $\dfrac{1}{2^{12}}$

C) $\dfrac{1}{2^4}$ D) 1

57) Simplify to only positive exponents.

$$\frac{6^5}{6 \cdot 6^{-9} \cdot 6^0}$$

A) $\dfrac{1}{6^{14}}$ B) 6^2

C) 6^{13} D) $\dfrac{1}{6^2}$

58) Simplify to only positive exponents.

$$\frac{9^7 \cdot 9^{-4}}{9^5}$$

A) 9^{10} B) $\dfrac{1}{9^2}$

C) $\dfrac{1}{9^4}$ D) 9^{21}

59) Simplify to only positive exponents.

$$\left(2^{-4}\right)^{-1} \cdot \left(2^2\right)^{-3}$$

A) $\dfrac{1}{2^2}$ B) 2^4

C) 2^2 D) 2^{15}

60) Simplify to only positive exponents.

$$\left(2^{-2}\right)^4 \cdot 2^2$$

A) $\dfrac{1}{2^6}$ B) 2

C) $\dfrac{1}{2^{44}}$ D) 2^6

Grade 8

Vol 1 Week 8 Assessment 1

61) Simplify to only positive exponents.

$$2^4 \cdot (2^4)^3$$

A) 1 B) 2^{16}

C) 2^{13} D) 2^6

62) Simplify to only positive exponents.

$$\frac{(2^{-2})^2}{2^2}$$

A) $\frac{1}{2}$ B) $\frac{1}{2^4}$

C) $\frac{1}{2^6}$ D) 2^{11}

63) Simplify to only positive exponents.

$$\left(\frac{2^4}{2}\right)^2$$

A) 2^{11} B) 2^3

C) $\frac{1}{2^2}$ D) 2^6

64) Simplify to only positive exponents.

$$\frac{2^{-4}}{(2^{-4})^3}$$

A) 2 B) $\frac{1}{2^{16}}$

C) $\frac{1}{2^{13}}$ D) 2^8

65) Simplify to only positive exponents.

$$\frac{(2^{-3})^4}{2^0 \cdot 2^{-1}}$$

A) $\frac{1}{2^{45}}$ B) 2

C) $\frac{1}{2^6}$ D) $\frac{1}{2^{11}}$

66) Simplify to only positive exponents.

$$\frac{(2^{-2})^4}{2^0}$$

A) $\frac{1}{2^8}$ B) 1

C) 2^7 D) 2^8

67) Simplify to only positive exponents.

$$\frac{(2^3)^2}{(2^4)^2}$$

A) 2^3 B) $\frac{1}{2^{15}}$

C) $\frac{1}{2^2}$ D) 2^{18}

68) Simplify to only positive exponents.

$$\frac{7n^{-15} \cdot 2n^{-10}}{6n^{13} \cdot n}$$

A) $\frac{1}{42n^7}$ B) $\frac{7}{3n^{39}}$

C) $\frac{12n^8}{7}$ D) $\frac{11}{9n^{30}}$

Grade 8

Vol 1
Week 8
Assessment 1

69) Simplify to only positive exponents.

$$\left(\frac{2 \cdot 2^2}{2^3}\right)^3$$

A) $\dfrac{1}{2^4}$ B) 2^4

C) 1   D) 2^{19}

70) Simplify to only positive exponents.

$$\frac{2}{(2^0 \cdot 2^3)^{-1}}$$

A) $\dfrac{1}{2^{10}}$ B) $\dfrac{1}{2^8}$

C) $\dfrac{1}{2^{17}}$ D) 2^4

71) Simplify to only positive exponents.

$$\left((m^{-19})^{-25} \cdot m^{-6}\right)^{10}$$

A) $\dfrac{1}{m^{478}}$ B) $2m^{266}$

C) m^{4690} D) $\dfrac{1}{m^{75}}$

72) Simplify to only positive exponents.

$$(2n^{13})^{-2} \cdot n^0$$

A) $\dfrac{4}{n^{209}}$ B) $4096\, n^{27}$

C) $\dfrac{1}{4n^{26}}$ D) $\dfrac{1}{32n^{11}}$

Grade 8

Vol 1
Week 9
Slopes

1) Find the slope of the line from the graph.

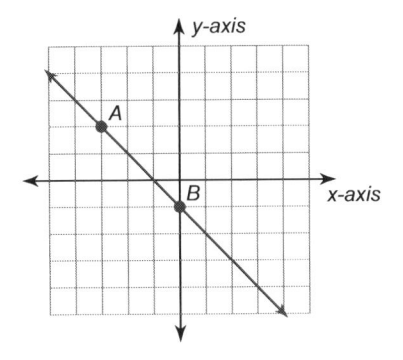

A) 1 B) −1

C) $-\dfrac{1}{3}$ D) $\dfrac{1}{3}$

2) Find the slope of the line from the graph.

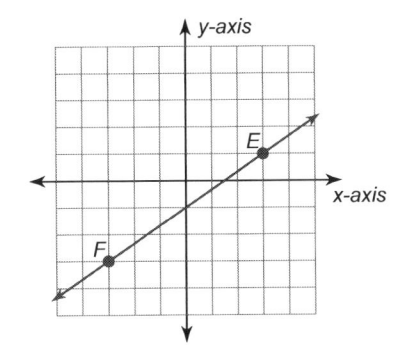

A) $\dfrac{3}{2}$ B) $-\dfrac{2}{3}$

C) $\dfrac{2}{3}$ D) $-\dfrac{3}{2}$

3) Find the slope of the line from the graph.

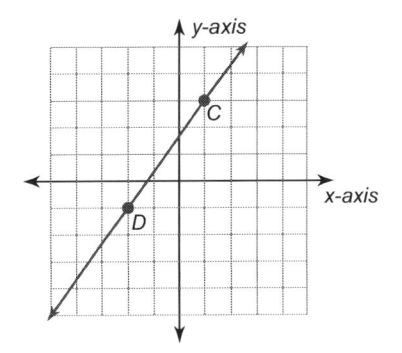

A) $\dfrac{3}{4}$ B) $\dfrac{4}{3}$

C) $-\dfrac{3}{4}$ D) $-\dfrac{4}{3}$

4) Find the slope of the line from the graph.

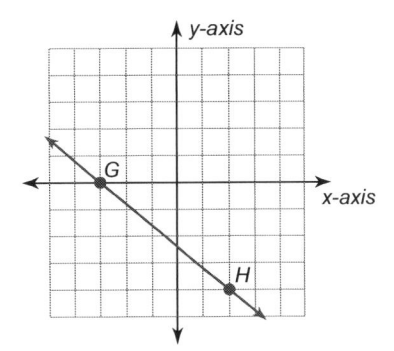

A) $\dfrac{5}{4}$ B) $\dfrac{4}{5}$

C) $-\dfrac{5}{4}$ D) $-\dfrac{4}{5}$

Grade 8

Vol 1
Week 9
Slopes

5) Find the slope of the line from the graph.

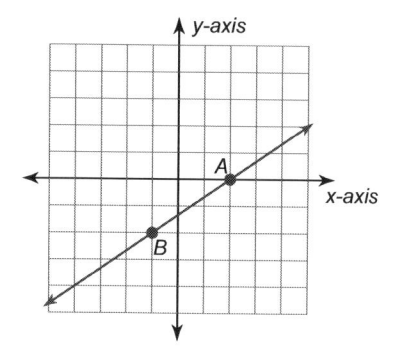

A) $\dfrac{3}{2}$ B) $\dfrac{2}{3}$

C) $-\dfrac{2}{3}$ D) $-\dfrac{3}{2}$

6) Find the slope of the line from the graph.

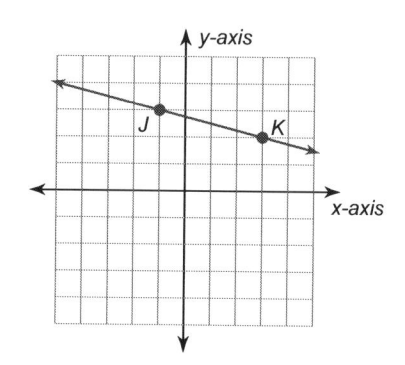

A) $-\dfrac{1}{4}$ B) 4

C) $\dfrac{1}{4}$ D) -4

7) Find the slope of the line from the graph.

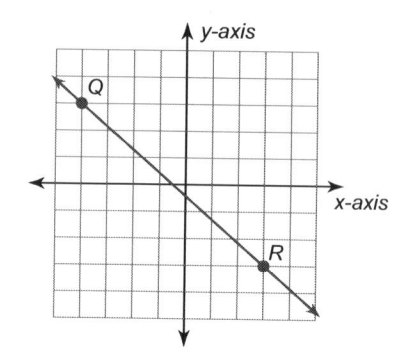

A) $\dfrac{7}{6}$ B) $-\dfrac{7}{6}$

C) $\dfrac{6}{7}$ D) $-\dfrac{6}{7}$

8) Find the slope of the line from the graph.

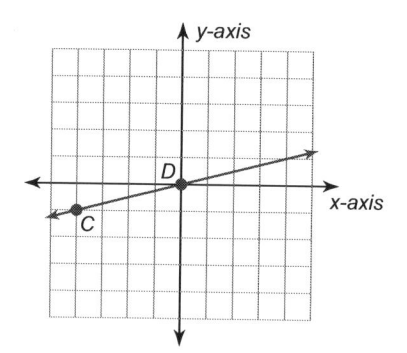

A) $\dfrac{1}{4}$ B) -4

C) 4   D) $-\dfrac{1}{4}$

©All rights reserved-Math-Knots LLC., VA-USA

www.math-knots.com | www.a4ace.com

Grade 8

Vol 1 Week 9 Slopes

9) Find the slope of the line from the graph.

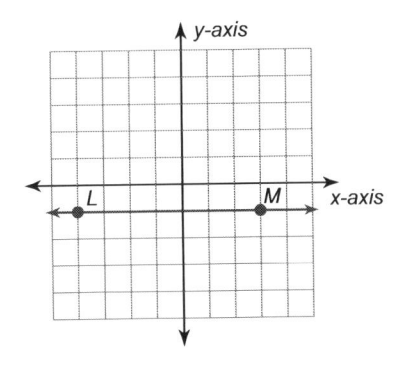

A) 0 B) $\dfrac{2}{3}$

C) $-\dfrac{2}{3}$ D) Undefined

11) Find the slope of the line from the graph.

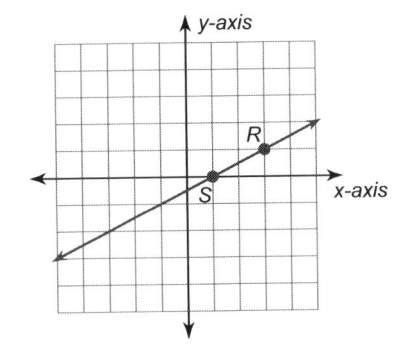

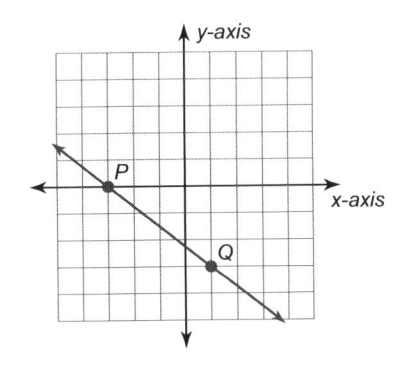

A) $-\dfrac{3}{4}$ B) $\dfrac{4}{3}$

C) $-\dfrac{4}{3}$ D) $\dfrac{3}{4}$

10) Find the slope of the line from the graph.

A) $\dfrac{1}{2}$ B) −2

C) $-\dfrac{1}{2}$ D) 2

12) Find the slope of the line from the graph.

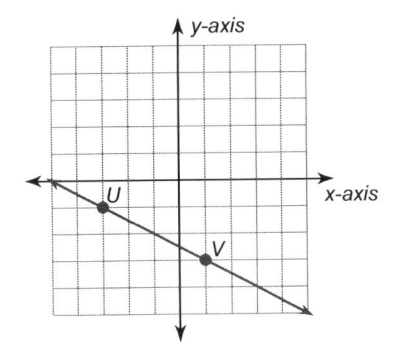

A) 2 B) $-\dfrac{1}{2}$

C) $\dfrac{1}{2}$ D) −2

Grade 8

Vol 1 Week 9 Slopes

13) Find the slope of the line passing through the points P (-19, -19) and Q (12, -12)

A) $\dfrac{7}{31}$ B) $-\dfrac{7}{31}$

C) $\dfrac{31}{7}$ D) $-\dfrac{31}{7}$

14) Find the slope of the line passing through the points P (-20, 3) and Q (-17, 0)

A) $\dfrac{5}{4}$ B) 1

C) $-\dfrac{5}{4}$ D) -1

15) Find the slope of the line passing through the points P (9, -11) and Q (10, 14)

A) $-\dfrac{1}{25}$ B) 25

C) $\dfrac{1}{25}$ D) -25

16) Find the slope of the line passing through the points P (-12, 17) and Q (1, 9)

A) $-\dfrac{8}{13}$ B) $-\dfrac{13}{8}$

C) $\dfrac{8}{13}$ D) $\dfrac{13}{8}$

17) Find the slope of the line passing through the points P (14, -10) and Q (-3, -6)

A) $\dfrac{17}{4}$ B) $-\dfrac{4}{17}$

C) $-\dfrac{17}{4}$ D) $\dfrac{4}{17}$

18) Find the slope of the line passing through the points P (-19, -19) and Q (12, -12)

A) $\dfrac{16}{27}$ B) $-\dfrac{27}{16}$

C) $\dfrac{27}{16}$ D) $-\dfrac{16}{27}$

19) Find the slope of the line passing through the points P (3, -9) and Q (-17, 0)

A) $-\dfrac{9}{20}$ B) $-\dfrac{20}{9}$

C) $\dfrac{20}{9}$ D) $\dfrac{9}{20}$

20) Find the slope of the line passing through the points P (-10, -11) and Q (19, 2)

A) $-\dfrac{13}{29}$ B) $-\dfrac{29}{13}$

C) $\dfrac{13}{29}$ D) $\dfrac{29}{13}$

Grade 8

Vol 1 Week 9 Slopes

21) Find the slope of the line passing through the points P (13, 6) and Q (1, -3)

A) $\dfrac{4}{3}$ B) $-\dfrac{4}{3}$

C) $\dfrac{3}{4}$ D) $-\dfrac{3}{4}$

22) Find the slope of the line passing through the points P (3, 19) and Q (-18, -15)

A) $\dfrac{34}{21}$ B) $-\dfrac{21}{34}$

C) $\dfrac{21}{34}$ D) $-\dfrac{34}{21}$

23) Find the slope of the line passing through the points P (19, -17) and Q (-18, -13)

A) $\dfrac{37}{4}$ B) $\dfrac{4}{37}$

C) $-\dfrac{37}{4}$ D) $-\dfrac{4}{37}$

24) Find the slope of the line passing through the points P (-3, 9) and Q (-6, -1)

A) $\dfrac{10}{3}$ B) $-\dfrac{10}{3}$

C) $-\dfrac{3}{10}$ D) $\dfrac{3}{10}$

25) Find the slope of the line passing through the points P (-18, 15) and Q (14, -16)

A) $-\dfrac{31}{32}$ B) $-\dfrac{32}{31}$

C) $\dfrac{31}{32}$ D) $\dfrac{32}{31}$

26) Find the slope of the line passing through the points P (-14, -10) and Q (-1, -3)

A) $-\dfrac{7}{13}$ B) $-\dfrac{13}{7}$

C) $\dfrac{7}{13}$ D) $\dfrac{13}{7}$

27) Find the missing coordinate given the points on the straight line with a slope of $-\dfrac{6}{5}$ passing through the points A(3, 13) and B(x, 1)

A) 10   B) 13

C) 11   D) 0

28) Find the missing coordinate given the points on the straight line with a slope of -1 passing through the points E(-1, y) and F(-3, -5)

A) −10   B) 13

C) 3   D) −7

Grade 8

Vol 1 Week 9 Slopes

29) Find the slope of the line passing through the points P (-10, -7) and Q (-4, -18)

   A) $\dfrac{6}{11}$ B) $-\dfrac{11}{6}$

   C) $\dfrac{11}{6}$ D) $-\dfrac{6}{11}$

30) Find the slope of the line passing through the points P (14, 13) and Q (-5, 18)

   A) $\dfrac{5}{19}$ B) $\dfrac{19}{5}$

   C) $-\dfrac{5}{19}$ D) $-\dfrac{19}{5}$

31) Find the missing coordinate given the points on the straight line with a slope of $-\dfrac{6}{7}$ passing through the points

   A(8,y) and B(-6,4)

   A) 14 B) -3

   C) -8 D) 5

32) Find the missing coordinate given the points on the straight line with a slope of 24 passing through the points C(-6,13) and D(x,-11)

   A) 12 B) -8

   C) -7 D) -6

33) Find the slope of the line passing through the points P (-2, -5) and Q (-7, 3)

   A) $\dfrac{8}{5}$ B) $\dfrac{5}{8}$

   C) $-\dfrac{5}{8}$ D) $-\dfrac{8}{5}$

34) Find the slope of the line passing through the points P (-18, 14) and Q (11, -12)

   A) $\dfrac{29}{26}$ B) $-\dfrac{29}{26}$

   C) $-\dfrac{26}{29}$ D) $\dfrac{26}{29}$

35) Find the missing coordinate given the points on the straight line with a slope of -16 passing through the points A(7,y) and B(8,-3)

   A) -12 B) 4

   C) -2 D) 13

36) Find the missing coordinate given the points on the straight line with a slope of $-\dfrac{1}{3}$ passing through the points

   C(x,1) and D(-2,-2)

   A) -3 B) -11

   C) -9 D) 5

Grade 8 — Vol 1, Week 9, Slopes

37) Find the missing coordinate given the points on the straight line with a slope of $-\dfrac{11}{3}$ passing through the points

J(7,-10) and K(1,y)

A) -9   B) -14

C) 12   D) -3

38) Find the missing coordinate given the points on the straight line with a slope of $-\dfrac{16}{5}$ passing through the points

P(-1,y) and Q(-6,6)

A) -10   B) 2

C) 7   D) -14

39) Find the missing coordinate given the points on the straight line with a slope of $\dfrac{11}{8}$ passing through the points

R(-2,y) and S(6,9)

A) 4   B) -2

C) 11   D) 0

40) Find the missing coordinate given the points on the straight line with a slope of -1 passing through the points
S(-4,-7) and T(x,3)

A) 2   B) 8

C) 0   D) -14

41) Find the missing coordinate given the points on the straight line with a slope of $-\dfrac{7}{4}$ passing through the points

U(0,6) and V(x,-8)

A) 12   B) 8

C) 14   D) -4

42) Find the missing coordinate given the points on the straight line with a slope of -1 passing through the points
F(4,-13) and G(x,-11)

A) -14   B) 5

C) 2   D) 14

43) Find the missing coordinate given the points on the straight line with a slope of -11 passing through the points
U(-8,11) and V(x,0)

A) -14   B) 8

C) 10   D) -7

44) Find the missing coordinate given the points on the straight line with a slope of 4 passing through the points
S(-9,4) and T(-11,y)

A) 5   B) -4

C) 0   D) 11

Grade 8

Vol 1 Week 9 Slopes

45) Find the missing coordinate given the points on the straight line with a slope of $-\dfrac{5}{2}$ passing through the points

A(x,-6) and B(-13,4)

A) 14 B) -9
C) 1 D) 0

46) Find the missing coordinate given the points on the straight line with a slope of $\dfrac{10}{3}$ passing through the points

C(x,-6) and D(7,14)

A) 1 B) -1
C) 6 D) 2

47) Find the slope of the straight line given below.

$y = -3x + 4$

A) $\dfrac{1}{3}$ B) -3
C) $-\dfrac{1}{3}$ D) 3

48) Find the slope of the straight line given below.

$y = -\dfrac{5}{3}x + 4$

A) $\dfrac{5}{3}$ B) $\dfrac{3}{5}$
C) $-\dfrac{5}{3}$ D) $-\dfrac{3}{5}$

49) Find the slope of the straight line given below.

$y = -\dfrac{1}{2}x + 3$

A) $-\dfrac{1}{2}$ B) 2
C) -2 D) $\dfrac{1}{2}$

50) Find the slope of the straight line given below.

$x = -3$

A) $\dfrac{1}{5}$ B) Undefined
C) $-\dfrac{1}{5}$ D) 0

51) Find the slope of the straight line given below.

$y = -4x$

A) 4 B) $-\dfrac{1}{4}$
C) -4 D) $\dfrac{1}{4}$

52) Find the slope of the straight line given below.

$y = x + 3$

A) 3 B) -1
C) 1 D) -3

©All rights reserved-Math-Knots LLC., VA-USA

www.math-knots.com | www.a4ace.com

Grade 8

Vol 1 Week 9 Slopes

53) Find the slope of the straight line given below.

$$y = \frac{3}{2}x - 5$$

A) $\frac{2}{3}$ B) $\frac{3}{2}$

C) $-\frac{2}{3}$ D) $-\frac{3}{2}$

54) Find the slope of the straight line given below.

$$y = x + 1$$

A) −1 B) −5

C) 5 D) 1

55) Find the slope of the straight line given below.

$$y = x + 2$$

A) $-\frac{4}{5}$ B) $\frac{4}{5}$

C) −1 D) 1

56) Find the slope of the straight line given below.

$$y = -x - 4$$

A) −1 B) 1

C) −2 D) 2

57) Find the slope of the straight line given below.

$$y = -x - 2$$

A) $\frac{1}{5}$ B) $-\frac{1}{5}$

C) 1 D) −1

58) Find the slope of the straight line given below.

$$y = -9x + 4$$

A) $-\frac{1}{9}$ B) −9

C) 9 D) $\frac{1}{9}$

59) Find the slope of the straight line given below.

$$y = -2x + 4$$

A) −2 B) $\frac{1}{2}$

C) $-\frac{1}{2}$ D) 2

60) Find the slope of the straight line given below.

$$y = -\frac{3}{2}x - 4$$

A) $\frac{3}{2}$ B) $-\frac{3}{2}$

C) $-\frac{2}{3}$ D) $\frac{2}{3}$

©All rights reserved-Math-Knots LLC., VA-USA   www.math-knots.com | www.a4ace.com

Grade 8

Vol 1 Week 9 Slopes

61) Find the slope of the straight line given below.

$$y = 4x + 3$$

A) −4 B) −$\frac{1}{4}$
C) $\frac{1}{4}$ D) 4

62) Find the slope of the straight line given below.

$$y = \frac{1}{4}x + 1$$

A) −4 B) −$\frac{1}{4}$
C) 4 D) $\frac{1}{4}$

63) Find the slope of the straight line given below.

$$y = \frac{4}{3}x$$

A) −$\frac{4}{3}$ B) $\frac{4}{3}$
C) −$\frac{3}{4}$ D) $\frac{3}{4}$

64) Find the slope of the straight line given below.

$$y = 7x - 3$$

A) 7 B) −$\frac{1}{7}$
C) $\frac{1}{7}$ D) −7

65) Find the slope of the straight line given below.

$$y = 7x + 4$$

A) $\frac{1}{7}$ B) −$\frac{1}{7}$
C) −7 D) 7

66) Find the slope of the straight line given below.

$$y = -x - 3$$

A) $\frac{3}{2}$ B) −1
C) −$\frac{3}{2}$ D) 1

67) Find the slope of the straight line given below.

$$x + 5y = 15$$

A) −$\frac{1}{5}$ B) $\frac{1}{5}$
C) −5 D) 5

68) Find the slope of the straight line given below.

$$x + 3y = -15$$

A) 3 B) $\frac{1}{3}$
C) −3 D) −$\frac{1}{3}$

Grade 8

Vol 1 Week 9 Slopes

69) Find the slope of the straight line given below.

$$y = -x + 1$$

A) 1   B) −1

C) $\dfrac{3}{2}$ D) $-\dfrac{3}{2}$

70) Find the slope of the straight line given below.

$$x - y = 2$$

A) 1   B) −1

C) $\dfrac{5}{4}$ D) $-\dfrac{5}{4}$

71) Find the slope of the straight line given below.

$$y = -3$$

A) Undefined   B) 0

C) $\dfrac{3}{4}$ D) $-\dfrac{3}{4}$

72) Find the slope of the straight line given below.

$$2x + y = 4$$

A) $\dfrac{1}{2}$ B) 2

C) $-\dfrac{1}{2}$ D) −2

73) Find the slope of the straight line given below.

$$9x - y = -5$$

A) 9   B) −9

C) $-\dfrac{1}{9}$ D) $\dfrac{1}{9}$

74) Find the slope of the straight line given below.

$$3x - 2y = -2$$

A) $-\dfrac{2}{3}$ B) $\dfrac{2}{3}$

C) $-\dfrac{3}{2}$ D) $\dfrac{3}{2}$

75) Find the slope of the straight line given below.

$$x - 4y = -16$$

A) 4   B) $-\dfrac{1}{4}$

C) $\dfrac{1}{4}$ D) −4

76) Find the slope of the straight line given below.

$$4x - 3y = -15$$

A) $\dfrac{4}{3}$ B) $\dfrac{3}{4}$

C) $-\dfrac{4}{3}$ D) $-\dfrac{3}{4}$

Grade 8

**Vol 1
Week 9
Slopes**

77) Find the slope of the straight line given below.

$$3x + 2y = -6$$

A) $-\dfrac{3}{2}$ B) $\dfrac{3}{2}$

C) $-\dfrac{2}{3}$ D) $\dfrac{2}{3}$

78) Find the slope of the straight line given below.

$$5x + 2y = 8$$

A) $-\dfrac{5}{2}$ B) $-\dfrac{2}{5}$

C) $\dfrac{2}{5}$ D) $\dfrac{5}{2}$

79) Find the slope of the straight line given below.

$$6x + 5y = -20$$

A) $\dfrac{6}{5}$ B) $-\dfrac{6}{5}$

C) $-\dfrac{5}{6}$ D) $\dfrac{5}{6}$

80) Find the slope of the straight line given below.

$$5x - 2y = -10$$

A) $-\dfrac{5}{2}$ B) $-\dfrac{2}{5}$

C) $\dfrac{2}{5}$ D) $\dfrac{5}{2}$

81) Find the slope of the straight line given below.

$$2x - 3y = 3$$

A) $\dfrac{2}{3}$ B) $-\dfrac{2}{3}$

C) $-\dfrac{3}{2}$ D) $\dfrac{3}{2}$

82) Find the slope of the straight line given below.

$$x + 4y = 4$$

A) -4 B) $-\dfrac{1}{4}$

C) 4 D) $\dfrac{1}{4}$

83) Which of the below straight line has the least slope?

A) $y = x - 14$ B) $y = 4x - 14$

C) $y = -4x - 14$ D) $y = -x - 14$

84) Which of the below straight line has the highest slope?

A) $y = \dfrac{5}{2}x + \dfrac{15}{2}$ B) $y = -\dfrac{15}{2}x - \dfrac{5}{2}$

C) $y = -\dfrac{5}{2}x + \dfrac{15}{2}$ D) $y = \dfrac{15}{2}x - \dfrac{5}{2}$

©All rights reserved-Math-Knots LLC., VA-USA   www.math-knots.com | www.a4ace.com

Grade 8

Vol 1 Week 9 Slopes

85) Find the slope of the straight line given below.

$$3x - y = 2$$

A) $\dfrac{1}{3}$ B) $-\dfrac{1}{3}$

C) -3 D) 3

86) Find the slope of the straight line given below.

$$5x + y = -5$$

A) $-\dfrac{1}{5}$ B) $\dfrac{1}{5}$

C) -5 D) 5

87) Find the slope of the straight line given below.

$$x + y = -4$$

A) -1 B) -2

C) 1 D) 2

88) Which of the below straight line has the least slope?

A) $y = -\dfrac{4}{3}x - \dfrac{2}{3}$

B) $y = 2$

C) $y = \dfrac{1}{3}x - \dfrac{2}{3}$

D) $y = \dfrac{4}{3}x - \dfrac{2}{3}$

89) Find the slope of the straight line given below.

$$7x - 2y = 6$$

A) $\dfrac{7}{2}$ B) $\dfrac{2}{7}$

C) $-\dfrac{2}{7}$ D) $-\dfrac{7}{2}$

90) Find the slope of the straight line given below.

$$x + y = 0$$

A) 1 B) -1

C) $-\dfrac{5}{2}$ D) $\dfrac{5}{2}$

91) Which of the below straight line has the highest slope?

A) $y = 2x + \dfrac{1}{2}$ B) $y = 4$

C) $y = -\dfrac{1}{2}x + 2$ D) $y = \dfrac{1}{2}x + 2$

92) Which of the below straight line has the highest slope?

A) $y = x + \dfrac{1}{5}$

B) $y = \dfrac{1}{5}x - 2$

C) $y = -2x + \dfrac{1}{5}$

D) $y = -\dfrac{2}{5}x + \dfrac{1}{5}$

Grade 8

Vol 1
Week 9
Slopes

93) Which of the below straight line has the least slope?

A) $y = \frac{1}{3}x + 4$

B) $y = -\frac{7}{3}x + 4$

C) $y = -\frac{1}{3}x + 4$

D) $y = \frac{7}{3}x + 4$

94) Which of the below straight line has the highest slope?

A) $y = -3x + 4$ B) $y = -4x + 4$

C) $y = x + 4$ D) $y = 4x - 3$

95) Which of the below straight line has the least slope?

A) $y = \frac{3}{2}x + 2$ B) $y = -2x + \frac{3}{2}$

C) $y = \frac{3}{2}x - 2$ D) $y = 2x + \frac{3}{2}$

96) Which of the below straight line has the highest slope?

A) $y = 4x + \frac{3}{4}$ B) $y = -\frac{1}{4}x + 4$

C) $y = \frac{3}{4}x + 4$ D) $y = -\frac{5}{4}x + 4$

97) Which of the below straight line has the least slope?

A) $y = -\frac{1}{7}x + \frac{1}{7}$

B) $y = \frac{1}{7}x + \frac{1}{7}$

C) $y = -\frac{5}{7}x + \frac{1}{7}$

D) $y = \frac{1}{7}x - \frac{1}{7}$

98) Which of the below straight line has the highest slope?

A) $y = -23x + 5$ B) $y = -101x - 100$

C) $y = 21x - 7$ D) $-3y = 0$

99) Which of the below straight line has the least slope?

A) $x - y = 1$ B) $x - y = -4$

C) $x + y = -4$ D) $x - y = -1$

Grade 8

Vol 1 Week 9 Slopes

100) Which of the below straight line has the least slope?

   A) $4x + 7y = -13$

   B) $x + 7y = 13$

   C) $13x - 7y = -4$

   D) $4x - 7y = -13$

101) Which of the below straight line has the least slope?

   A) $y = 2x - 1$ B) $y = 2x + 1$

   C) $y = x + 2$ D) $y = -x + 2$

102) Which of the below straight line has the least slope?

   A) $y = -\frac{2}{5}x - \frac{4}{5}$ B) $y = -\frac{4}{5}$

   C) $y = -9x - 7$ D) $y = -2x + 5$

103) Which of the below straight line has the least slope?

   A) $2x + y = 5$ B) $16x + 5y = -2$

   C) $5x - y = 2$ D) $2x + y = 0$

104) Which of the below straight line has the highest slope?

   A) $x - y = -4$

   B) $x + y = -4$

   C) $4x - 5y = -1$

   D) $4x + y = -5$

105) Which of the below straight line has the least slope?

   A) $3x + 2y = -4$

   B) $3x + 4y = -2$

   C) $3x - 2y = 4$

   D) $3x - 4y = 2$

106) Which of the below straight line has the highest slope?

   A) $2x + 7y = 8$

   B) $2x + 7y = 7$

   C) $7x + 2y = -8$

   D) $2x - 7y = -7$

Grade 8

Vol 1 Week 9 Slopes

107) Which of the below straight line has the least slope?

A) $x - y = 3$ B) $x + 3y = 1$
C) $x - y = 1$ D) $x - y = -1$

108) Which of the below straight line has the least slope?

A) $2x - 3y = -14$
B) $x + 3y = 14$
C) $2x + 3y = -14$
D) $x - 3y = 14$

109) Which of the below straight line has the least slope?

A) $2x + y = -7$ B) $7x + y = -2$
C) $4x - y = 7$ D) $2x - y = -7$

110) Which of the below straight line has the least slope?

A) $x - 2y = 4$ B) $2x + 2y = 1$
C) $2x + y = -4$ D) $2x - y = -4$

111) Which of the below straight line has the least slope?

A) $2x + y = 5$ B) $-y = 4$
C) $2x + y = -4$ D) $y = 0$

112) Find the function plotted below.

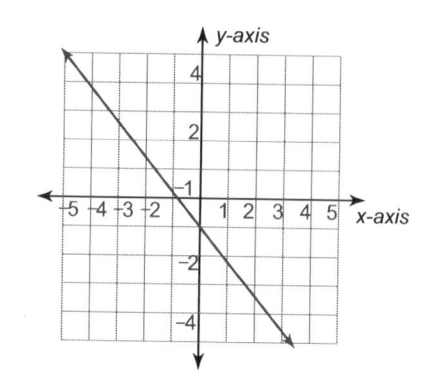

A) $6x + 5y = 5$
B) $6x - 5y = 5$
C) $6x - 5y = -5$
D) $6x + 5y = -5$

113) Find the function plotted below.

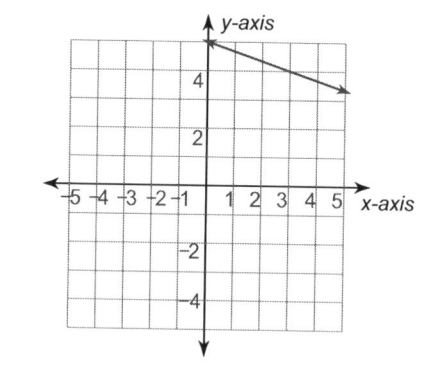

A) $x + 3y = 15$
B) $x + 3y = -15$
C) $x - 3y = -15$
D) $x - 3y = 9$

Grade 8

Vol 1 Week 9 Slopes

114) Find the function plotted below.

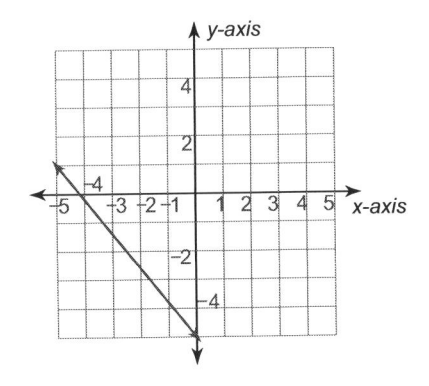

A) $y = -\dfrac{6}{5}x - 5$ B) $y = -x - 5$

C) $y = -\dfrac{2}{5}x - 5$ D) $y = x - 5$

116) Find the function plotted below.

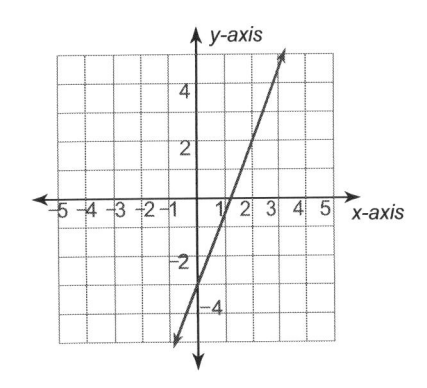

A) $y = 3x - 3$ B) $y = -4x - 3$

C) $y = -x - 3$ D) $y = \dfrac{5}{2}x - 3$

115) Find the function plotted below.

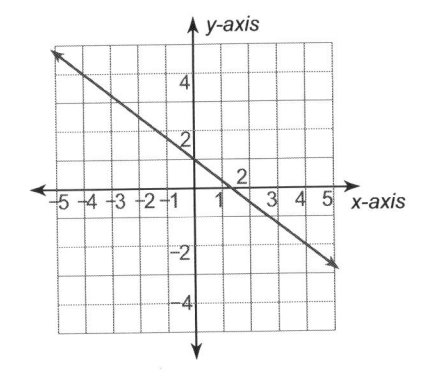

A) $y = \dfrac{5}{4}x + 1$ B) $y = -\dfrac{5}{4}x + 1$

C) $y = \dfrac{3}{4}x + 1$ D) $y = -\dfrac{3}{4}x + 1$

117) Find the function plotted below.

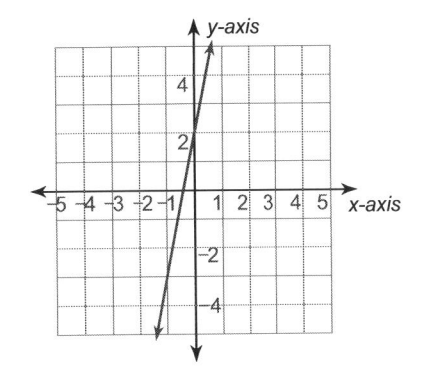

A) $y = -2x + 2$ B) $y = 5x + 2$

C) $y = 3x + 2$ D) $y = 2x + 2$

Grade 8

Vol 1
Week 9
Slopes

118) Find the function plotted below.

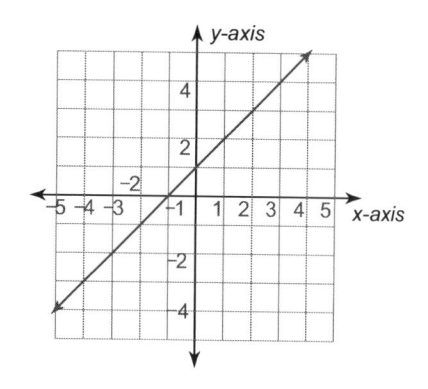

A) $y = 5x + 1$ B) $y = x + 1$

C) $y = -x + 1$ D) $y = 4x + 1$

120) Find the function plotted below.

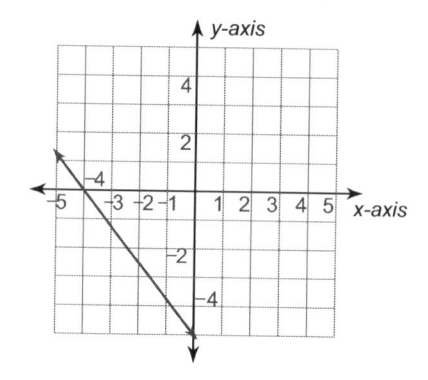

A) $y = -5x + \dfrac{5}{4}$ B) $y = -\dfrac{1}{2}x + \dfrac{5}{4}$

C) $y = \dfrac{5}{4}x - 5$ D) $y = -\dfrac{5}{4}x - 5$

119) Find the function plotted below.

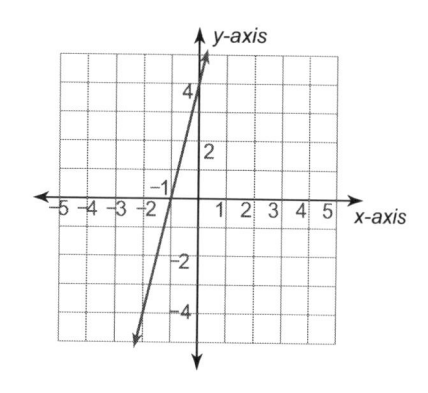

A) $y = 4x + 4$ B) $y = -2x + 4$

C) $y = 3x + 4$ D) $y = -4x + 4$

121) Find the function plotted below.

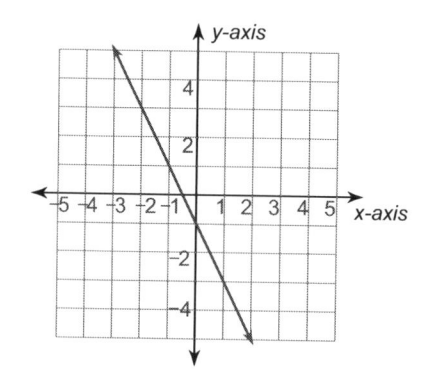

A) $y = -2x - 1$ B) $y = x + 2$

C) $y = -x + 2$ D) $y = 2x - 1$

Grade 8

Vol 1 Week 9 Slopes

122) Find the function plotted below.

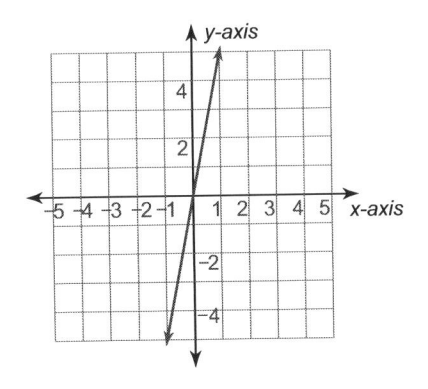

A) $y = 3x$ B) $y = 5x$

C) $y = -5x$ D) $y = -3x$

123) Find the function plotted below.

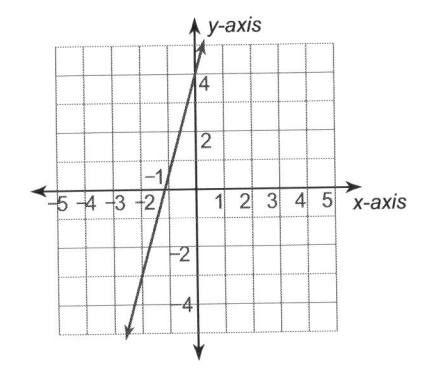

A) $y = -\dfrac{5}{2}x + \dfrac{7}{2}$

B) $y = 4x + \dfrac{7}{2}$

C) $y = \dfrac{7}{2}x + 4$

D) $y = -4x + \dfrac{7}{2}$

124) Find the function plotted below.

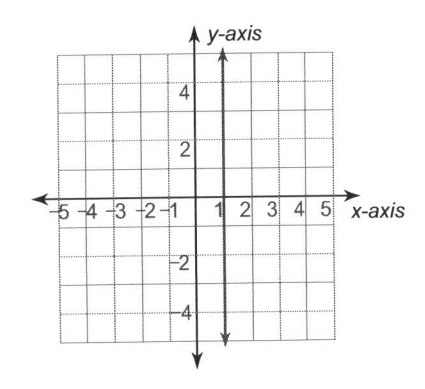

A) $x = 1$ B) $x = -1$

C) $y = -1$ D) $y = x$

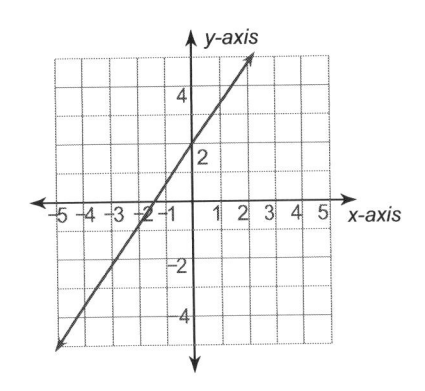

125) Find the function plotted below.

A) $y = \dfrac{4}{5}x + 2$

B) $y = -\dfrac{4}{5}x + 2$

C) $y = -\dfrac{2}{5}x + 2$

D) $y = \dfrac{7}{5}x + 2$

©All rights reserved-Math-Knots LLC., VA-USA

126) Find the function plotted below.

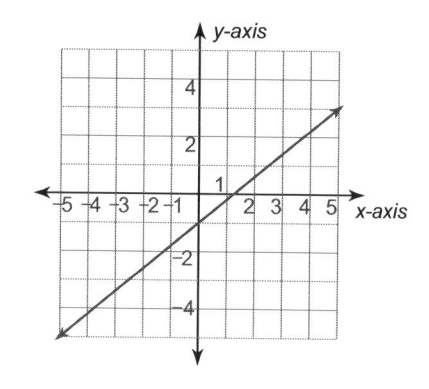

A) $y = -\dfrac{4}{5}x - 1$ B) $y = \dfrac{3}{5}x - 1$

C) $y = -\dfrac{3}{5}x - 1$ D) $y = \dfrac{4}{5}x - 1$

127) Find the function plotted below.

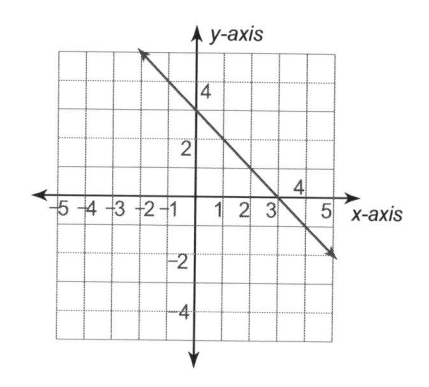

A) $y = -x + 3$ B) $y = 5x + 3$

C) $y = 3x - 1$ D) $y = 3x + 5$

128) Find the function plotted below.

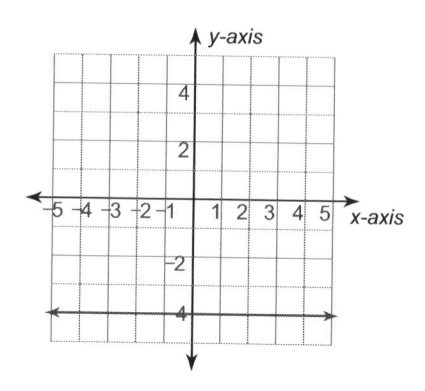

A) $y = -4x + 3$ B) $y = 4x + 3$

C) $y = 3x - 4$ D) $y = -4$

129) Find the function plotted below.

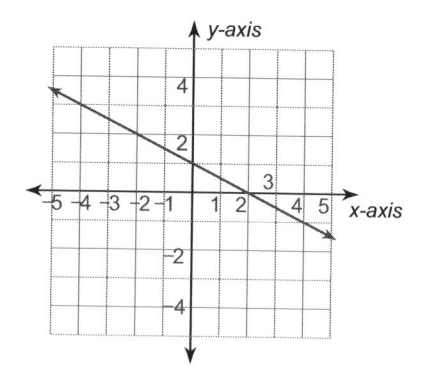

A) $y = -2x + 1$

B) $y = -\dfrac{1}{2}x + 1$

C) $y = 2x + 1$

D) $y = -5x + 1$

Grade 8

Vol 1 Week 9 Slopes

130) Find the function plotted below.

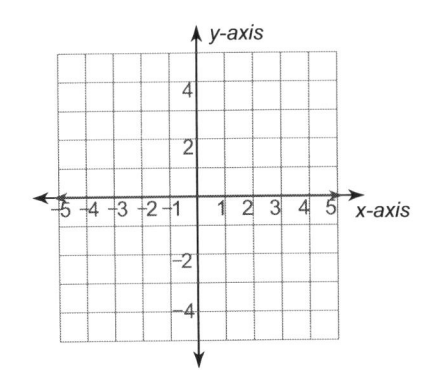

A) $y = -x$ B) $y = -\dfrac{1}{2}x$

C) $y = \dfrac{1}{2}x$ D) $y = 0$

132) Find the function plotted below.

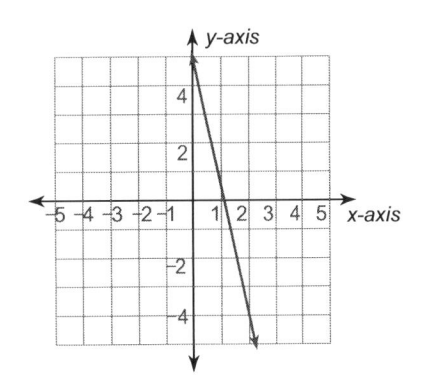

A) $y = 5x + 2$ B) $y = -\dfrac{9}{2}x + 5$

C) $y = 2x + 5$ D) $y = -2x + 5$

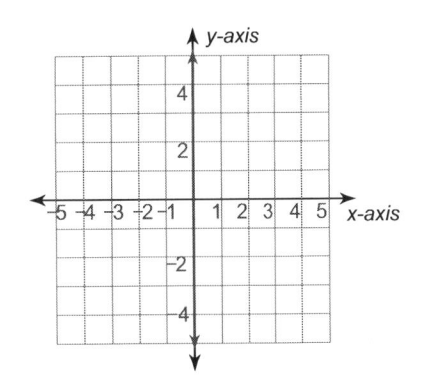

131) Find the function plotted below.

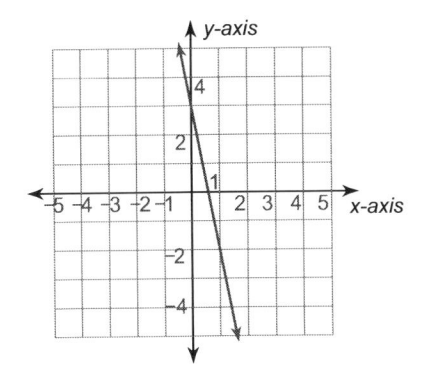

A) $y = x - 5$ B) $y = 3x - 5$

C) $y = -5x + 3$ D) $y = 5x + 3$

133) Find the function plotted below.

A) $y = \dfrac{1}{5}x + \dfrac{1}{5}$ B) $y = 0$

C) $y = \dfrac{1}{5}$ D) $x = 0$

Grade 8

Vol 1
Week 9
Slopes

134) Find the function plotted below.

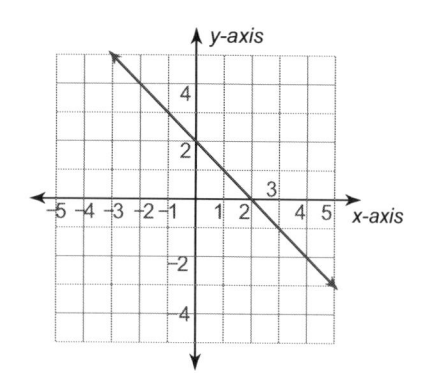

A) $y = 5x + 2$ B) $y = -x + 2$

C) $y = 2x - 1$ D) $y = x + 2$

135) Find the function plotted below.

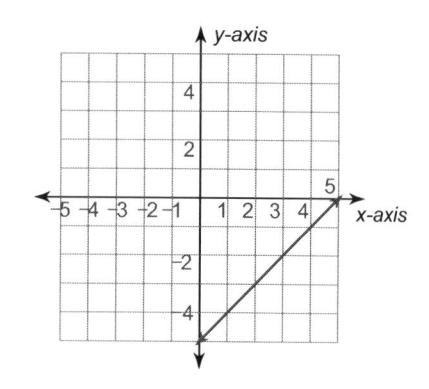

A) $y = -x - 5$ B) $y = -5x - 5$

C) $y = 5x - 5$ D) $y = x - 5$

136) Find the function plotted below.

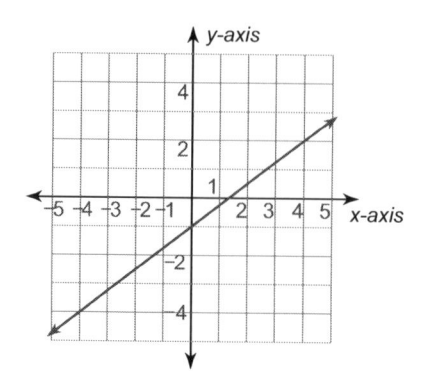

A) $y = \frac{3}{4}x - 1$ B) $y = -\frac{1}{2}x - 1$

C) $y = -\frac{3}{4}x - 1$ D) $y = \frac{5}{2}x - 1$

137) Find the function plotted below.

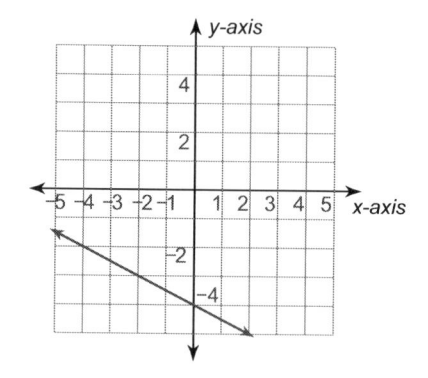

A) $y = -\frac{1}{2}x - 4$

B) $y = -4x - \frac{1}{2}$

C) $y = -x - 4$

D) $y = -4x - 1$

Grade 8

Vol 1
Week 9
Slopes

138) Find the function plotted below.

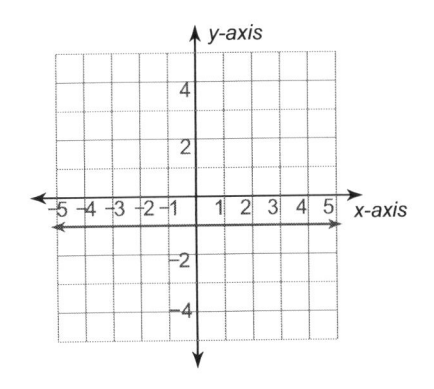

A) $y = -1$ B) $x = 1$

C) $y = -x$ D) $y = 4x$

139) Find the function plotted below.

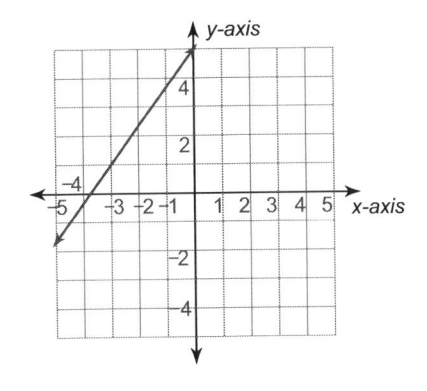

A) $3x + 4y = 15$

B) $4x + 3y = -15$

C) $3x - 15y = -4$

D) $4x - 3y = -15$

140) Find the function plotted below.

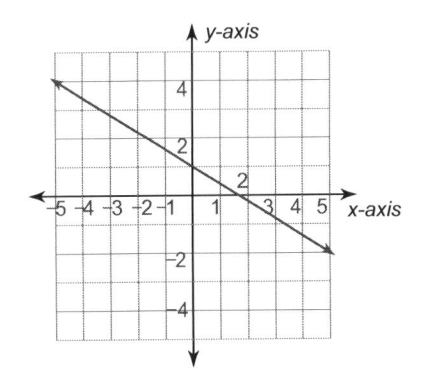

A) $3x + 5y = 5$

B) $3x - 5y = -2$

C) $5x - 3y = 2$

D) $5x + 5y = -2$

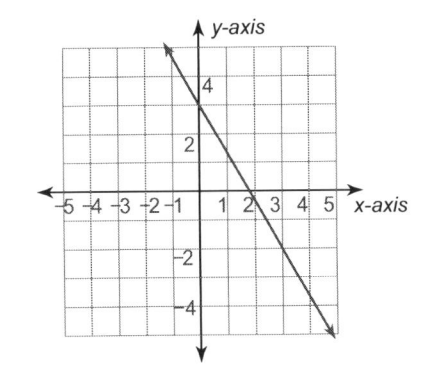

141) Find the function plotted below.

A) $5x + 3y = 9$

B) $3x - 5y = 9$

C) $3x + 5y = 9$

D) $5x - 3y = -9$

Grade 8

Vol 1 Week 9 Slopes

142) Find the function plotted below.

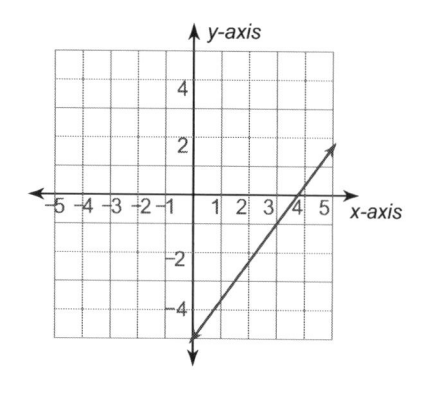

A) $x - 3y = -3$

B) $4x - 3y = 3$

C) $4x + 3y = -3$

D) $4x - 3y = 15$

143) Find the function plotted below.

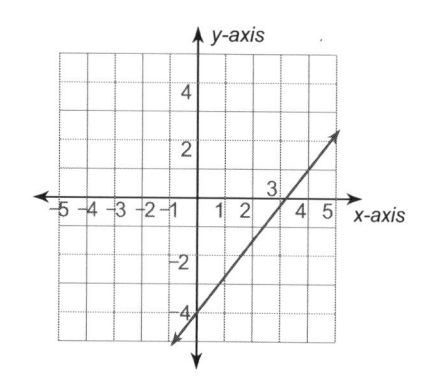

A) $-y = -1$ B) $5x - 4y = 16$

C) $x + y = 0$ D) $5x + 4y = 8$

144) Find the function plotted below.

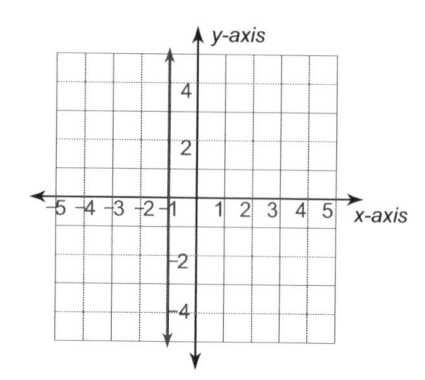

A) $x = -1$ B) $4x = 0$

C) $x = 0$ D) $3x = 0$

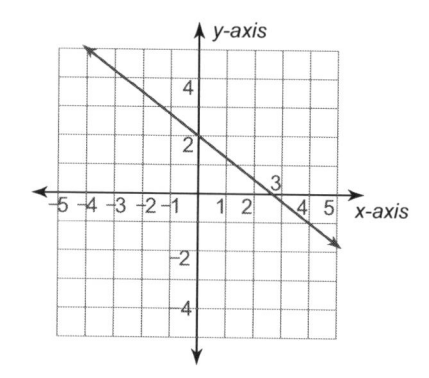

145) Find the function plotted below.

A) $y = 4$ B) $3x + 4y = 8$

C) $x - 4y = 8$ D) $-y = 5$

Grade 8

Vol 1
Week 9
Slopes

146) Find the function plotted below.

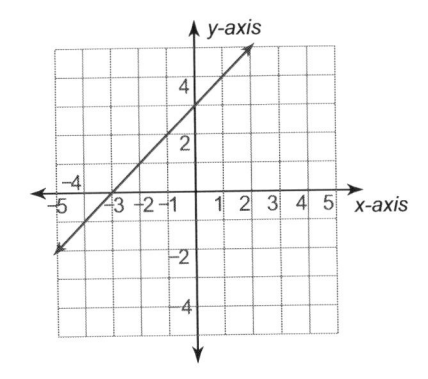

A) $3x - y = 5$

B) $x - y = -3$

C) $4x + 5y = 1$

D) $4x + 5y = -1$

147) Find the function plotted below.

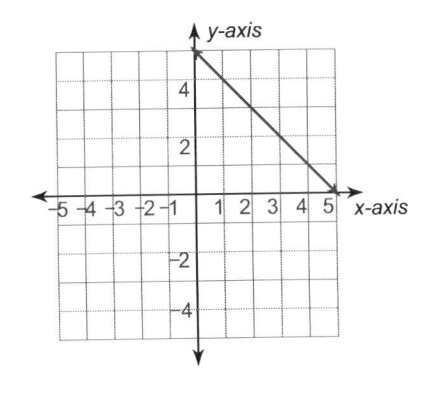

A) $4x + 5y = -1$

B) $4x - y = -1$

C) $4x + 5y = -20$

D) $x + y = 5$

148) Find the function plotted below.

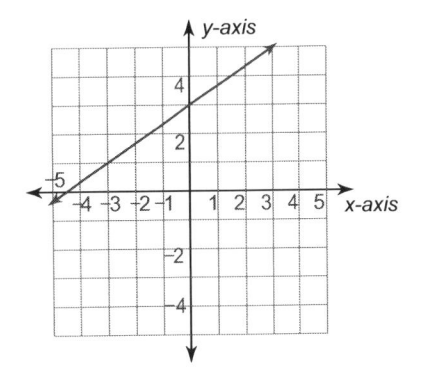

A) $2x - 3y = -9$ B) $x - 3y = 9$

C) $3x + y = -6$ D) $6x + y = 3$

149) Find the function plotted below.

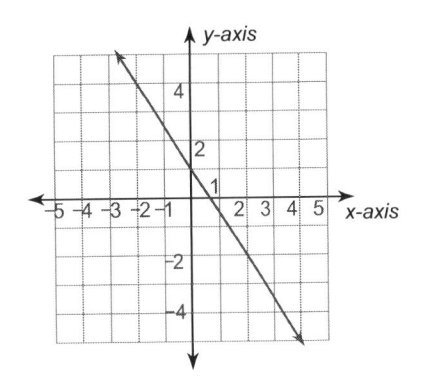

A) $3x + 2y = 2$

B) $3x + 2y = -2$

C) $2x - y = 2$

D) $3x - 2y = -2$

Grade 8

Vol 1
Week 9
Slopes

150) Find the function plotted below.

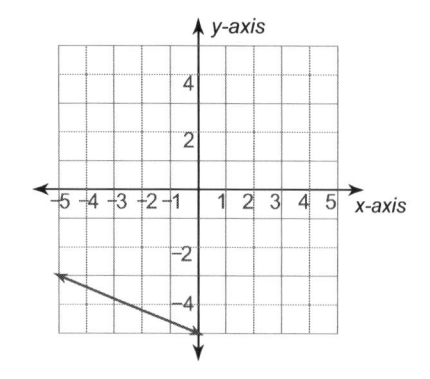

A) $x - 5y = -5$

B) $25x - 5y = 3$

C) $2x + 5y = -25$

D) $5x - 3y = -25$

151) Find the function plotted below.

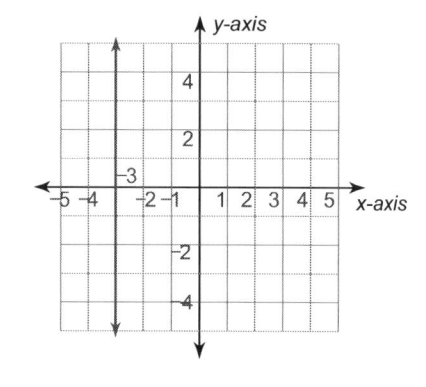

A) $-y = 0$ B) $x + 4y = 0$

C) $y = 0$ D) $x = -3$

152) Find the function plotted below.

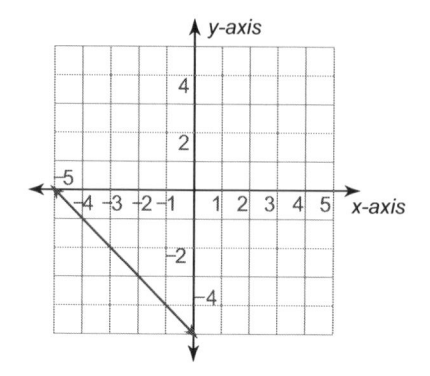

A) $x + y = -5$ B) $2x - y = 2$

C) $4x - y = 5$ D) $2x + y = 2$

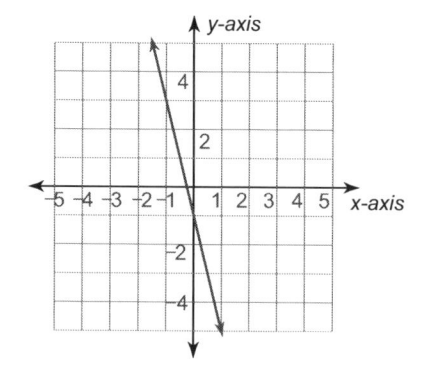

153) Find the function plotted below.

A) $4x + y = -1$ B) $5x - y = -1$

C) $x - y = -5$ D) $2x + y = -1$

Grade 8

Vol 1
Week 9
Slopes

154) Find the function plotted below.

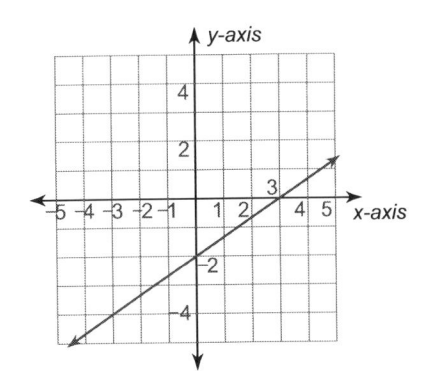

A) $2x - 3y = 6$

B) $2x + 3y = -6$

C) $2x - 3y = -15$

D) $2x - 3y = 15$

155) Find the function plotted below.

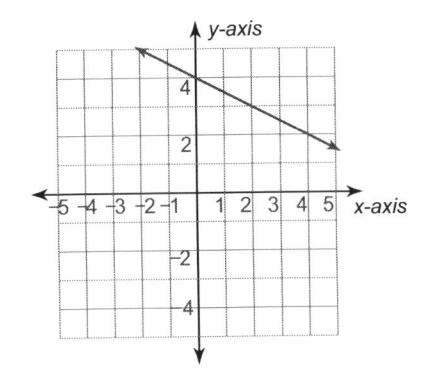

A) $x + 2y = 10$

B) $x - 4y = -16$

C) $x + 4y = -1$

D) $x + 2y = 8$

156) Find the function plotted below.

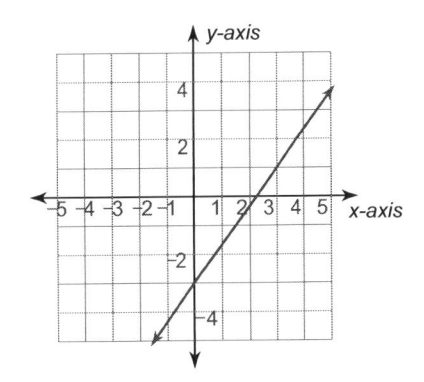

A) $x - y = 2$ B) $4x - 3y = 9$

C) $4x - 3y = 6$ D) $x + y = -2$

157) Find the function plotted below.

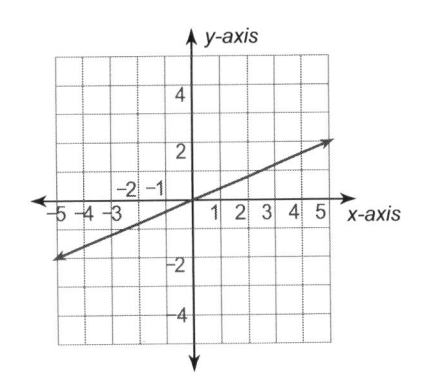

A) $2x + 15y = -5$

B) $2x + 15y = -30$

C) $2x - 5y = -15$

D) $2x - 5y = 0$

Grade 8

**Vol 1
Week 9
Slopes**

158) Find the function plotted below.

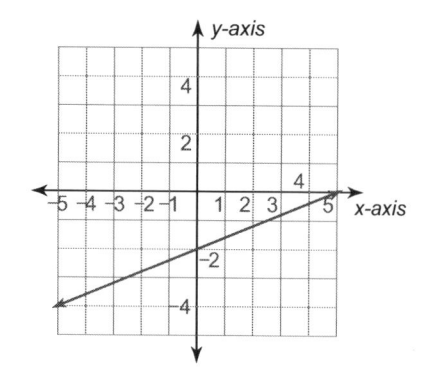

A) $5x - 2y = -10$

B) $10x - 2y = 5$

C) $2x - 5y = 10$

D) $10x + 2y = -5$

159) Find the function plotted below.

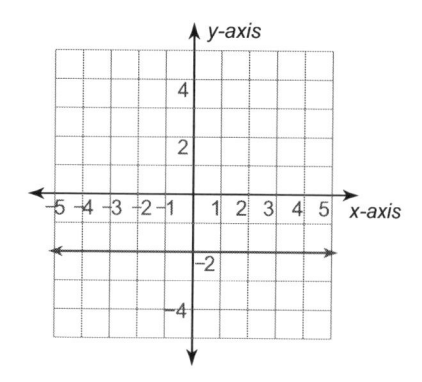

A) $y = 2$ B) $y = -2$

C) $x = 2$ D) $x = -2$

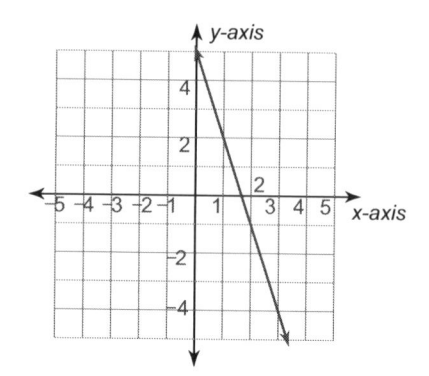

160) Find the function plotted below.

A) $5x + 3y = -1$

B) $3x + y = 5$

C) $5x - y = 3$

D) $5x - 3y = 1$

Grade 8

Vol 1 Week 10 Linear equations

1) Find the equation of the straight line passing through P(2,5) and parallel to $y = -\dfrac{7}{3}x + 2$

A) $y = -\dfrac{7}{3}x + \dfrac{29}{3}$ B) $y = \dfrac{29}{3}x + \dfrac{7}{3}$

C) $y = -\dfrac{29}{3}x + \dfrac{7}{3}$ D) $y = \dfrac{7}{3}x + \dfrac{29}{3}$

2) Find the equation of the straight line passing through P(-3,1) and parallel to $y = -\dfrac{2}{3}x + 1$

A) $y = -\dfrac{5}{3}x - 1$ B) $y = -\dfrac{2}{3}x - 1$

C) $y = -\dfrac{2}{3}x - \dfrac{5}{3}$ D) $y = -x - \dfrac{5}{3}$

3) Find the equation of the straight line passing through P(-3,4) and parallel to $x = 0$

A) $y = -\dfrac{3}{4}x$ B) $x = -3$

C) $y = -\dfrac{5}{4}x - \dfrac{3}{4}$ D) $y = -\dfrac{3}{4}$

4) Find the equation of the straight line passing through P(-3,1) and parallel to $y = \dfrac{1}{3}x - 4$

A) $y = -2x + \dfrac{1}{3}$ B) $y = -x + \dfrac{1}{3}$

C) $y = \dfrac{1}{3}x + 2$ D) $y = 2x + \dfrac{1}{3}$

5) Find the equation of the straight line passing through P(-5,2) and parallel to $y = \dfrac{1}{5}x - 1$

A) $y = -\dfrac{3}{5}x + \dfrac{1}{5}$ B) $y = -x + \dfrac{1}{5}$

C) $y = 3x + \dfrac{1}{5}$ D) $y = \dfrac{1}{5}x + 3$

6) Find the equation of the straight line passing through P(3,3) and parallel to $y = \dfrac{7}{3}x - 1$

A) $y = -x - 4$ B) $y = 4x - 1$

C) $y = \dfrac{7}{3}x - 4$ D) $y = -4x - 1$

Grade 8

Vol 1 Week 10 Linear equations

7) Find the equation of the straight line passing through P(-2,0) and parallel to $y = -\frac{2}{5}x + 1$

A) $y = \frac{2}{5}x - \frac{4}{5}$
B) $y = -\frac{2}{5}x - \frac{4}{5}$
C) $y = -\frac{1}{5}x - \frac{4}{5}$
D) $y = \frac{4}{5}x - \frac{4}{5}$

8) Find the equation of the straight line passing through P(-2,-5) and parallel to $y = 2x + 3$

A) $y = -2x - 1$
B) $y = -x + 2$
C) $y = -x - 2$
D) $y = 2x - 1$

9) Find the equation of the straight line passing through P(5,3) and parallel to $y = \frac{3}{5}x - 5$

A) $y = -\frac{3}{5}x$
B) $y = \frac{3}{5}$
C) $y = \frac{3}{5}x$
D) $x = 3$

10) Find the equation of the straight line passing through P(-1,1) and parallel to $y = x + 3$

A) $y = 3x + 2$
B) $y = x + 2$
C) $y = 2x + 2$
D) $y = -3x + 2$

11) Find the equation of the straight line passing through P(-1,5) and parallel to $y = -7x + 5$

A) $y = -2x - 7$
B) $y = -7x - 2$
C) $y = -5x - 2$
D) $y = 5x - 2$

12) Find the equation of the straight line passing through P(-5,-5) and parallel to $y = x + 3$

A) $x = -1$
B) $y = 1$
C) $x = 1$
D) $y = x$

©All rights reserved-Math-Knots LLC., VA-USA
www.math-knots.com | www.a4ace.com

Grade 8

Vol 1 Week 10 Linear equations

13) Write the standard form of the equation of the straight line passing through P(-2,2) and parallel to $y = -\dfrac{3}{7}x + 5$

A) $3x + 7y = 8$ B) $14x + 7y = 2$

C) $14x + 7y = -2$ D) $14x - 7y = -2$

14) Write the standard form of the equation of the straight line passing through P(5,5) and parallel to $y = \dfrac{9}{5}x - 5$

A) $9x - 5y = -20$

B) $5x + y = -1$

C) $9x - 5y = 20$

D) $25x - 5y = -4$

15) Write the standard form of the equation of the straight line passing through P(-1,3) and parallel to $x = 0$

A) $x + y = -2$ B) $x + y = 0$

C) $x = -1$ D) $x - y = 0$

16) Write the standard form of the equation of the straight line passing through P(4,-3) and parallel to $y = -\dfrac{3}{2}x - 4$

A) $2x + 3y = 1$ B) $3x + 2y = 6$

C) $3x - 2y = 6$ D) $3x - 2y = 10$

17) Write the standard form of the equation of the straight line passing through P(1,-1) and parallel to $y = x + 2$

A) $x - 4y = -1$ B) $x + y = 2$

C) $x - y = 2$ D) $2x + y = 1$

18) Write the standard form of the equation of the straight line passing through P(-2,2) and parallel to $y = \dfrac{1}{2}x$

A) $2x - y = -6$ B) $x + 4y = 0$

C) $x - 2y = 6$ D) $x - 2y = -6$

Grade 8

Vol 1 Week 10 Linear equations

19) Write the standard form of the equation of the straight line passing through P(-2,-3) and parallel to $y = \frac{2}{3}x + 5$

A) $2x - 3y = 5$ B) $5x - 3y = 0$

C) $5x - 3y = -2$ D) $5x - 3y = -9$

20) Write the standard form of the equation of the straight line passing through P(-3,-1) and parallel to $y = -\frac{1}{2}x + 2$

A) $5x + 2y = -5$ B) $4x - 2y = -5$

C) $x + 2y = -5$ D) $2x - 2y = 5$

21) Write the standard form of the equation of the straight line passing through P(4,-1) and parallel to $y = \frac{3}{2}x + 1$

A) $3x + 2y = -10$ B) $3x - 2y = -10$

C) $3x - 2y = 10$ D) $3x + 2y = -2$

22) Write the standard form of the equation of the straight line passing through P(5,4) and parallel to $y = \frac{4}{5}x + 5$

A) $4x - 5y = 0$ B) $4x - 3y = 0$

C) $5x - 4y = 0$ D) $3x + 4y = 0$

23) Write the standard form of the equation of the straight line passing through P(-2,4) and parallel to $y = -\frac{9}{2}x + 4$

A) $x + 2y = -4$ B) $9x - 2y = 8$

C) $9x + 2y = -10$ D) $9x - 2y = -10$

24) Write the standard form of the equation of the straight line passing through P(-4,-5) and parallel to $y = \frac{3}{4}x - 3$

A) $3x - 4y = 8$ B) $x - 4y = 8$

C) $x + 4y = -8$ D) $3x + 4y = 8$

Grade 8

Vol 1 Week 10 Linear equations

25) Write the slope intercept form of the equation of the straight line passing through P(4,-1) and perpendicular to y = 2x + 1

A) $y = \frac{1}{2}x$ B) $y = \frac{1}{2}$

C) $y = -2x + 1$ D) $y = -\frac{1}{2}x$

26) Write the slope intercept form of the equation of the straight line passing through P(4,-2) and perpendicular to $y = -\frac{5}{2}x - 2$

A) $y = \frac{18}{5}x + 1$ B) $y = x - \frac{18}{5}$

C) $y = \frac{2}{5}x - \frac{18}{5}$ D) $y = -\frac{18}{5}x + 1$

27) Write the slope intercept form of the equation of the straight line passing through P(1,4) and perpendicular to y = -1

A) $y = -\frac{1}{2}$ B) $x = 1$

C) $x = -1$ D) $x = 2$

28) Write the slope intercept form of the equation of the straight line passing through P(2,-1) and perpendicular to y = x + 5

A) $y = -2x + 1$ B) $y = 4x + 1$

C) $y = x + 1$ D) $y = -x + 1$

29) Write the slope intercept form of the equation of the straight line passing through P(3,2) and perpendicular to $y = \frac{3}{2}x - 3$

A) $y = \frac{2}{3}x + 4$ B) $y = -\frac{2}{3}x + 4$

C) $y = -4x + \frac{2}{3}$ D) $y = 4x + \frac{2}{3}$

30) Write the slope intercept form of the equation of the straight line passing through P(-3,-1) and perpendicular to y = 4

A) $x = -3$ B) $x = 3$

C) $x = -5$ D) $x = 5$

Grade 8

Vol 1 Week 10 Linear equations

31) Write the slope intercept form of the equation of the straight line passing through P(-1,0) and perpendicular to $y = \frac{1}{4}x - 3$

A) $y = 3x - 4$ B) $y = -4x - 4$

C) $y = 5x - 4$ D) $y = -3x - 4$

32) Write the slope intercept form of the equation of the straight line passing through P(-5,1) and perpendicular to $y = x + 4$

A) $y = -4x - 4$ B) $y = 5x - 4$

C) $y = -x - 4$ D) $y = x - 4$

33) Write the slope intercept form of the equation of the straight line passing through P(2,-1) and perpendicular to $y = \frac{5}{2}x - 3$

A) $y = -\frac{1}{5}x + \frac{2}{5}$ B) $y = \frac{2}{5}x - \frac{1}{5}$

C) $y = -\frac{2}{5}x - \frac{1}{5}$ D) $y = -x + \frac{2}{5}$

34) Write the slope intercept form of the equation of the straight line passing through P(-3,-5) and perpendicular to $y = -\frac{1}{3}x - 5$

A) $y = -4x - 3$ B) $y = -3x + 4$

C) $y = 3x + 4$ D) $y = 4x - 3$

35) Write the slope intercept form of the equation of the straight line passing through P(5,-5) and perpendicular to $y = \frac{5}{4}x + 3$

A) $y = \frac{4}{5}x - 1$ B) $y = -x - \frac{4}{5}$

C) $y = -\frac{4}{5}x - 1$ D) $y = -x + \frac{4}{5}$

36) Write the slope intercept form of the equation of the straight line passing through P(1,3) and perpendicular to $y = \frac{1}{8}x + 4$

A) $y = 8x + 5$ B) $y = 2x + 5$

C) $y = -8x + 5$ D) $y = x + 5$

Grade 8

Vol 1 — Week 10 — Linear equations

37) Write the standard form of the equation of the straight line passing through P(-3,-4) and perpendicular to x = 0

A) $4x - y = 4$ B) $4x - y = -4$

C) $y = -4$ D) $2x - y = 4$

38) Write the standard form of the equation of the straight line passing through P(4,0) and perpendicular to y = x

A) $x - 5y = 0$ B) $x = -5$

C) $5x = 2$ D) $x + y = 4$

39) Write the standard form of the equation of the straight line passing through P(-1,3) and perpendicular to $y = \frac{1}{6}x + 2$

A) $6x + y = -3$ B) $3x - y = -6$

C) $3x + y = -6$ D) $3x - y = 6$

40) Write the standard form of the equation of the straight line passing through P(5,-1) and perpendicular to $y = -\frac{9}{4}x\ 5$

A) $4x - 9y = 29$ B) $4x - 29y = 9$

C) $29x + 4y = -4$ D) $87x - 29y = -4$

41) Write the standard form of the equation of the straight line passing through P(3,-1) and perpendicular to y = -2x + 2

A) $x - 2y = 5$ B) $5x + y = 1$

C) $x + 2y = -5$ D) $5x - y = 2$

42) Write the standard form of the equation of the straight line passing through P(3,-2) and perpendicular to $y = \frac{3}{5}x$

A) $5x + 3y = -9$

B) $5x + 3y = 9$

C) $5x - 3y = 9$

D) $5x - 3y = -3$

Grade 8

Vol 1
Week 10
Linear equations

43) Write the standard form of the equation of the straight line passing through P(-1,4) and perpendicular to $y = \dfrac{1}{9}x + 3$

A) $2x + y = 5$
B) $9x + y = -5$
C) $2x - y = -5$
D) $9x + y = 5$

44) Write the standard form of the equation of the straight line passing through P(4,5) and perpendicular to $y = -\dfrac{1}{6}x$

A) $2x + 2y = -1$
B) $19x - 2y = 1$
C) $6x - y = 19$
D) $19x - 2y = -1$

45) Write the standard form of the equation of the straight line passing through P(2,2) and perpendicular to $y = -x + 3$

A) $x - y = 0$
B) $3x - y = -1$
C) $x + y = -2$
D) $3x + y = 1$

46) Write the standard form of the equation of the straight line passing through P(-4,5) and perpendicular to $y = \dfrac{8}{3}x - 2$

A) $3x - 8y = -24$
B) $3x + 8y = 28$
C) $24x + 8y = -1$
D) $3x + 8y = 24$

47) Write the standard form of the equation of the straight line passing through P(-1,-5) and perpendicular to $y = -\dfrac{1}{6}x - 4$

A) $6x - y = 1$
B) $6x - y = -1$
C) $x + y = 6$
D) $x - y = -3$

48) Write the standard form of the equation of the straight line passing through P(4,5) and perpendicular to $y = -\dfrac{4}{3}x - 1$

A) $3x - 2y = 12$
B) $3x - 4y = -8$
C) $2x - 3y = -12$
D) $4x + 3y = -12$

Grade 8

Vol 1
Week 10
Linear equations

49) Find the value of m that satisfies the below equation

$163 = -3m - 5(1 + 5m)$

A) −2   B) −6
C) 12   D) −12

50) Find the value of v that satisfies the below equation

$140 = -7(4v - 4)$

A) −4   B) 4
C) 0   D) −13

51) Find the value of x that satisfies the below equation

$8(x + 5) = 104$

A) −14   B) No solution.
C) −10   D) 8

52) Find the value of m that satisfies the below equation

$6(6m - 4) = 264$

A) −6   B) 4
C) 0   D) 8

53) Find the value of n that satisfies the below equation

$266 = 7(5n + 8)$

A) 6   B) −15
C) 15   D) All real numbers

54) Find the value of a that satisfies the below equation

$95 = -5(5a + 1)$

A) −4   B) 9
C) 7   D) −3

55) Find the value of x that satisfies the below equation

$4x - 7(1 + 5x) = -162$

A) −2   B) 5
C) No solution.   D) −4

56) Find the value of p that satisfies the below equation

$-8(1 + 8p) = 184$

A) −1   B) −7
C) All real numbers   D) −3

57) Find the value of a that satisfies the below equation

$2(2 + 6a) + 2a = 102$

A) 12   B) All real numbers
C) 7   D) −12

58) Find the value of r that satisfies the below equation

$88 = 4(4 + 3r)$

A) 9   B) 1
C) No solution.   D) 6

©All rights reserved-Math-Knots LLC., VA-USA   www.math-knots.com | www.a4ace.com

Grade 8

Vol 1 Week 10 Linear equations

59) Find the value of k that satisfies the below equation
$7(6 - 2k) = 84$

A) 5 B) −4
C) −3 D) 6

60) Find the value of x that satisfies the below equation
$298 = 2 + 8(-5x + 2)$

A) −4 B) 12
C) −7 D) 9

61) Find the value of n that satisfies the below equation
$-104 = 8(5n + 2)$

A) 3 B) −4
C) −3 D) No solution.

62) Find the value of p that satisfies the below equation
$-138 = -6(4p + 3)$

A) No solution. B) −15
C) −10 D) 5

63) Find the value of n that satisfies the below equation
$-8(-n + 5) = -88$

A) −4 B) −14
C) −6 D) No solution.

64) Find the value of x that satisfies the below equation
$8(7 + 8x) = -136$

A) −3 B) 1
C) −6 D) No solution.

65) Find the value of n that satisfies the below equation
$-140 = 7(4 - 4n) + 4n$

A) 7 B) −3
C) 5 D) −4

66) Find the value of n that satisfies the below equation
$3n + 7(5n - 8) = 172$

A) −10 B) 2
C) −2 D) 6

67) Find the value of r that satisfies the below equation
$-139 = -6 - 7(6r + 7)$

A) 2 B) All real numbers
C) −13 D) −16

68) Find the value of x that satisfies the below equation
$-85 = -5(1 - 2x)$

A) −8 B) No solution.
C) 14 D) 4

Grade 8

Vol 1 Week 10 Linear equations

69) Find the value of x that satisfies the below equation

$-92 = -5x - 4(x + 5)$

A) 0 B) 1
C) 16 D) 8

70) Find the value of x that satisfies the below equation

$7(4x - 1) + 4 = -171$

A) -6 B) -9
C) No solution. D) 3

71) Find the value of n that satisfies the below equation

$7(1 - 5n) = 217$

A) 3 B) -6
C) -12 D) -9

72) Find the value of x that satisfies the below equation

$6(6 + 8x) = -252$

A) -6 B) 8
C) 7 D) 2

73) Find the value of n that satisfies the below equation

$7(8 + 7n) = -287$

A) All real numbers B) 4
C) -6 D) -7

74) Find the value of x that satisfies the below equation

$5 - 8(1 - 8x) = -451$

A) 5 B) -12
C) -7 D) -5

75) Find the value of x that satisfies the below equation

$-84 = -4(1 + x) - 6x$

A) All real numbers B) 9
C) 8 D) -9

76) Find the value of n that satisfies the below equation

$-4(4n + 6) = -136$

A) 7 B) 11
C) All real numbers D) -12

77) Find the value of v that satisfies the below equation

$-8(4 + 3v) = 112$

A) -6 B) 0
C) -3 D) 9

78) Find the value of n that satisfies the below equation

$7(-4n + 8) + 7 = -161$

A) 4 B) 15
C) 8 D) -1

Grade 8

Vol 1 Week 10 Linear equations

79) Find the value of x that satisfies the below equation
$-6(2x+6) = -84$

A) 14 B) −1
C) 10 D) 4

80) Find the value of p that satisfies the below equation
$6(-8p-1) = -150$

A) 15 B) 3
C) −16 D) No solution.

81) Find the value of v that satisfies the below equation
$-136 = 4(5v-4)$

A) −6 B) 12
C) 8 D) −9

82) Find the value of n that satisfies the below equation
$6 + 7(n+6) = 97$

A) 15 B) 7
C) −10 D) −1

83) Find the value of x that satisfies the below equation
$-245 = -7(-4x+7)$

A) 16 B) −7
C) −12 D) 4

84) Find the value of x that satisfies the below equation
$84 = 4(3x+5) - 4x$

A) 10 B) 8
C) −15 D) No solution.

85) Find the value of x that satisfies the below equation
$118 = 2(-5+8x)$

A) −9 B) 8
C) −5 D) 0

86) Find the value of b that satisfies the below equation
$7(3b-3) = 147$

A) 8 B) No solution.
C) −16 D) −5

87) Find the value of a that satisfies the below equation
$-264 = 8(1+7a) + 8$

A) 9 B) −16
C) 3 D) −5

88) Find the value of k that satisfies the below equation
$-7(3k-2) = -91$

A) −11 B) All real numbers
C) 5 D) −6

Grade 8

Vol 1 Week 10 Linear equations

89) Find the value of m that satisfies the below equation

$2(3x-2)-6(x+6) = -40$

A) 14 B) 10
C) All real numbers D) 4

90) Find the value of x that satisfies the below equation

$-7(1-5x)-7(x+7) = 28$

A) 2 B) -4
C) 3 D) -13

91) Find the value of v that satisfies the below equation

$65 = 6(4v+8)+7(7-4v)$

A) All real numbers B) -5
C) 8 D) 13

92) Find the value of k that satisfies the below equation

$-48 = 3(5k-8)+4(-3k-6)$

A) 9 B) 1
C) -5 D) 0

93) Find the value of r that satisfies the below equation

$14 = 8(1-3r)-(7r-6)$

A) 2 B) -15
C) -2 D) 0

94) Find the value of x that satisfies the below equation

$42 = -6(1-x)-3(-2x+8)$

A) 0 B) 6
C) -9 D) -10

95) Find the value of x that satisfies the below equation

$-76 = -7(8+2x)+3(6x-8)$

A) -4 B) 0
C) -11 D) 1

96) Find the value of r that satisfies the below equation

$4 = 6(-8r-4)+7(r+4)$

A) 0 B) No solution.
C) 6 D) -2

97) Find the value of a that satisfies the below equation

$-63 = -(2+a)-7(-7-8a)$

A) No solution. B) -2
C) -10 D) -5

98) Find the value of a that satisfies the below equation

$-27 = -6(1+2a)-7(a+3)$

A) 3 B) 9
C) -6 D) 0

Grade 8

**Vol 1 Week 10
Linear equations**

99) Find the value of a that satisfies the below equation
$$-1 = 2(5a-2) - 2(1+5a)$$

A) No solution. B) 9
C) 16 D) 7

100) Find the value of a that satisfies the below equation
$$-7 = -4(a+1) + 3(-6a-1)$$

A) 2 B) 15
C) 14 D) 0

101) Find the value of a that satisfies the below equation
$$-4(a+8) - (a-3) = -39$$

A) 2 B) 8
C) 11 D) -4

102) Find the value of v that satisfies the below equation
$$3(-5+4v) - 5(-3v+5) = -67$$

A) 12 B) -12
C) All real numbers D) -1

103) Find the value of a that satisfies the below equation
$$-7(5a+8) + 8(a-5) = -42$$

A) -12 B) -2
C) -11 D) All real numbers

104) Find the value of n that satisfies the below equation
$$-40 = -3(-1+n) - 5(1+7n)$$

A) 1 B) 2
C) 7 D) -7

105) Find the value of r that satisfies the below equation
$$-(4r-4) - 7(1-3r) = -71$$

A) -2 B) -4
C) -9 D) 10

106) Find the value of p that satisfies the below equation
$$-22 = 3(2+p) - 4(p+8)$$

A) -11 B) -3
C) -4 D) 8

107) Find the value of b that satisfies the below equation
$$43 = 5(7b-6) - 8(-5+3b)$$

A) -12 B) 3
C) 4 D) 8

108) Find the value of b that satisfies the below equation
$$-44 = -2(7b-3) - (-4+4b)$$

A) -1 B) 3
C) 7 D) 8

Grade 8

Vol 1 Week 10 Linear equations

109) Find the value of x that satisfies the below equation
$$-52 = -2(x-1) + 2(1+5x)$$

A) −15 B) 1

C) −7 D) All real numbers

110) Find the value of n that satisfies the below equation
$$-35 = -3(n+5) - 5(4-8n)$$

A) −3 B) 0

C) No solution. D) 16

111) Find the value of v that satisfies the below equation
$$3(1+6v) + 5(6+5v) = -10$$

A) 3 B) −14

C) −1 D) 8

112) Find the value of x that satisfies the below equation
$$2(3x-6) - 6(-6+7x) = -48$$

A) 9 B) 11

C) −8 D) 2

113) Find the value of a that satisfies the below equation
$$43 = -(1+4a) + 4(1-a)$$

A) 8 B) −4

C) −5 D) −7

114) Find the value of p that satisfies the below equation
$$3(1-7p) - 6(8-2p) = 27$$

A) −8 B) All real numbers

C) 6 D) −3

115) Find the value of n that satisfies the below equation
$$-39 = -(-6n+1) - 4(4n-8)$$

A) −8 B) −10

C) −11 D) 7

116) Find the value of x that satisfies the below equation
$$50 = -4(-7x-4) + 2(-5x-1)$$

A) 10 B) −7

C) −4 D) 2

117) Find the value of r that satisfies the below equation
$$-6(3r+4) - 2(r+4) = -72$$

A) 2 B) −15

C) No solution. D) 16

118) Find the value of b that satisfies the below equation
$$42 = 3(b+6) - 3(b-8)$$

A) −15 B) No solution.

C) 7 D) All real numbers

©All rights reserved-Math-Knots LLC., VA-USA

Grade 8

Vol 1
Week 10
Linear equations

119) Find the value of x that satisfies the below equation
$6(6x+2) - 2(x-7) = -42$

A) 16 B) -2
C) -14 D) 3

120) Find the value of r that satisfies the below equation
$4(r+5) + 7(r-4) = -8$

A) All real numbers B) -8
C) 0 D) -3

121) Find the value of x that satisfies the below equation
$-17 = 4(3x-5) + 7(x-5)$

A) 4 B) 16
C) 2 D) -9

122) Find the value of a that satisfies the below equation
$-50 = -5(-6a+4) + 5(-6+a)$

A) -4 B) 11
C) 0 D) 4

123) Find the value of m that satisfies the below equation
$-3(2m-4) - 2(3m-1) = 38$

A) -2 B) 5
C) -5 D) 7

124) Find the value of a that satisfies the below equation
$7 = -6(a-8) + 3(2a-5)$

A) No solution. B) 10
C) 12 D) -4

125) Find the value of x that satisfies the below equation
$20 = 6(1+2x) - 7(-2-8x)$

A) 15 B) -3
C) All real numbers D) 0

126) Find the value of x that satisfies the below equation
$0 = 4(5+7x) - (x-7)$

A) -9 B) -1
C) 13 D) 9

127) Find the value of n that satisfies the below equation
$62 = -3(5-n) + 2(7-6n)$

A) 16 B) -6
C) -7 D) -11

128) Find the value of b that satisfies the below equation
$30 = 2(b-2) + 7(b+1)$

A) -13 B) -11
C) 3 D) 14

Grade 8

Vol 1 Week 10 — Linear equations

129) Find the value of r that satisfies the below equation
$8(r-7) = 2(r+2)$

A) 10 B) No solution.
C) -12 D) 8

130) Find the value of n that satisfies the below equation
$-1 + 3(-n+3) = 2(n+4) - 5n$

A) -12 B) -13
C) No solution. D) All real numbers

131) Find the value of n that satisfies the below equation
$-2n - 4(-1 - 3n) = 3(-6 + 4n)$

A) 11 B) 7
C) -1 D) 16

132) Find the value of x that satisfies the below equation
$-4(x-3) = -2(x+8)$

A) 14 B) -5
C) 10 D) -15

133) Find the value of x that satisfies the below equation
$2(x-8) + 3(8+4x) = 6x + 8x$

A) No solution. B) 4
C) 15 D) 9

134) Find the value of a that satisfies the below equation
$48 = -4(1 - 4a) - 4(a+8)$

A) No solution. B) 12
C) 7 D) 8

135) Find the value of v that satisfies the below equation
$28 = -5(7 - 7v) + 7(v-3)$

A) 4 B) 3
C) 2 D) -7

136) Find the value of x that satisfies the below equation
$-24 = 6(5x - 2) - 6(2 + 5x)$

A) All real numbers B) -10
C) 0 D) 2

137) Find the value of v that satisfies the below equation
$7(3v + 8) + 3(6v - 5) = 80$

A) -15 B) -10
C) 1 D) -13

138) Find the value of v that satisfies the below equation
$-24 = -4(v+7) + 2(v+5)$

A) 11 B) 3
C) -11 D) 6

Grade 8

Vol 1 Week 11 Linear equations

1) Find the value of a that satisfies the below equation

   $5m + 2m = 4(m + 5) + 2(m - 6)$

   A) −15   B) 8
   C) 14   D) −13

2) Find the value of a that satisfies the below equation

   $7(x + 7) = -7x - 7(2x + 5)$

   A) −3   B) 11
   C) −1   D) 1

3) Find the value of a that satisfies the below equation

   $-4(3p + 6) = 8(p - 8)$

   A) 15   B) 2
   C) 3   D) 12

4) Find the value of a that satisfies the below equation

   $-3(2 + 2n) = -6 - 4(n + 1)$

   A) −7   B) 1
   C) −6   D) 2

5) Find the value of a that satisfies the below equation

   $7(a + 2) + 4a = -2(1 - a) + 7$

   A) −1   B) 15
   C) −3   D) 8

6) Find the value of p that satisfies the below equation

   $-2(p - 5) = 2(4p + 2) + 6$

   A) −11   B) 0
   C) −6   D) 11

7) Find the value of m that satisfies the below equation

   $-5(3 + 2m) = 2(-5 - 6m) + 7m$

   A) −1   B) 14
   C) 7   D) 2

8) Find the value of b that satisfies the below equation

   $6(5b - 5) = -2(6b - 6)$

   A) 1   B) 9
   C) No solution.   D) 6

9) Find the value of b that satisfies the below equation

   $2b - 3b = 7(b - 2) - 2(-4 + 3b)$

   A) All real numbers   B) 4
   C) 3   D) 16

10) Find the value of v that satisfies the below equation

    $-6(v + 4) = 6(v + 2)$

    A) −3   B) 9
    C) −6   D) −5

Grade 8

Vol 1 Week 11 Linear equations

11) Find the value of n that satisfies the below equation
$-3(2+n)+2=7(2n-3)$

A) 11   B) −7
C) 1   D) −12

12) Find the value of x that satisfies the below equation
$-4(x-6)+8=-3(1-x)$

A) 7   B) −12
C) 5   D) All real numbers

13) Find the value of x that satisfies the below equation
$6(x+4)=-8(-x-1)+4$

A) 10   B) 6
C) 9   D) −6

14) Find the value of n that satisfies the below equation
$-6(-7+8n)=3(4-6n)$

A) 9   B) 13
C) 1   D) No solution.

15) Find the value of a that satisfies the below equation
$-2(-4a+3)=-5+7(a-1)$

A) 1   B) 7
C) −6   D) −14

16) Find the value of a that satisfies the below equation
$-2(7a-4)+3(-6+2a)=2a-5-8a+3$

A) −10 B) 3
C) −4   D) −3

17) Find the value of a that satisfies the below equation
$1+6n+7n-8=4(6n-1)-2(4+5n)$

A) 5   B) 7
C) −2   D) 15

18) Find the value of a that satisfies the below equation
$-2(5p+7)=6(3-7p)$

A) 11   B) 1
C) −15 D) −12

19) Find the value of a that satisfies the below equation
$2(4n+2)=4(n+2)$

A) 6   B) −9
C) 1   D) 10

20) Find the value of a that satisfies the below equation
$3(-3x+1)-7x=3(1+x)$

A) 0   B) 10
C) 13   D) No solution.

Grade 8

**Vol 1
Week 11
Linear equations**

21) Find the value of x that satisfies the below equation
$$-2(-5+x) = 4(2x-3) - 8$$

 A) −9 B) 14
 C) 3 D) No solution.

22) Find the value of x that satisfies the below equation
$$-(8+x) = -2x - (x+6)$$

 A) −15 B) 4
 C) 15 D) 1

23) Find the value of x that satisfies the below equation
$$2(1+m) - 3m = -4(m+3) + 4m$$

 A) 10 B) 14
 C) −3 D) −12

24) Find the value of x that satisfies the below equation
$$6(4+3k) = -3(k+6)$$

 A) No solution. B) 1
 C) −2 D) 0

25) Find the value of x that satisfies the below equation
$$3(n-1) = -4(n-8)$$

 A) 5 B) 10
 C) 11 D) 8

26) Find the value of x that satisfies the below equation
$$8x + 7(1-6x) = 7(x+1)$$

 A) −4 B) 0
 C) −1 D) −8

27) Find the value of r that satisfies the below equation
$$6r + 3(5r-2) = 6(r-1)$$

 A) −10 B) 3
 C) −9 D) 0

28) Find the value of n that satisfies the below equation
$$-(2n-1) + 7n = -8 - 4(1-n)$$

 A) 16 B) −13
 C) 5 D) −16

29) Find the value of r that satisfies the below equation
$$4(r+6) = 6r + 7(r-3)$$

 A) 12 B) All real numbers
 C) No solution. D) 5

30) Find the value of x that satisfies the below equation
$$-(-5-7x) = -(-7-5x)$$

 A) 2 B) −5
 C) 1 D) 8

Grade 8

Vol 1
Week 11
Linear equations

31) Find the value of x that satisfies the below equation
$3(4n + 7) = -(2n - 7)$

A) 13 B) −1
C) 10 D) 3

32) Find the value of x that satisfies the below equation
$-2(5a - 7) + a = -2(3a + 5)$

A) −13 B) −5
C) 8 D) 4

33) Find the value of x that satisfies the below equation
$6(3p + 3) = -6(-1 - 4p)$

A) 8 B) −10
C) 2 D) 1

34) Find the value of x that satisfies the below equation
$-2(x + 4) = -8(1 - 6x)$

A) 0 B) 7
C) −15 D) −2

35) Find the value of x that satisfies the below equation
$-8(1 + 2m) = 2(8 - 6m)$

A) 10 B) No solution.
C) −15 D) −6

36) Find the value of v that satisfies the below equation
$v - 3 + v - 1 = 7(5v + 5) + 3(-2v + 5)$

A) No solution. B) −2
C) 1 D) 9

37) Find the value of x that satisfies the below equation
$-7(1 - 5m) = 2(6 + 6m) + 4m$

A) 6 B) −9
C) 1 D) −12

38) Find the value of x that satisfies the below equation
$5(x - 6) + 1 = 5(x - 3) + 5$

A) 0 B) 6
C) No solution. D) 3

39) Find the value of x that satisfies the below equation
$5(m - 3) = 6(m - 2) + 1$

A) −11 B) 13
C) −4 D) All real numbers

40) Find the value of x that satisfies the below equation
$-8(1 - 3x) = 8(2x - 4)$

A) −4 B) −2
C) 9 D) −3

Grade 8

**Vol 1
Week 11
Linear equations**

41) Find the value of x that satisfies the below equation

$$\frac{624}{7} = -\frac{26}{7}\left(\frac{32}{7}x - \frac{8}{7}\right)$$

A) $-\frac{2}{3}$　　B) $\frac{7}{4}$

C) -5　　D) $-\frac{8}{7}$

42) Find the value of x that satisfies the below equation

$$\frac{34}{7}\left(\frac{23}{6}n + \frac{11}{3}\right) + \frac{13}{3}n = \frac{5942}{63}$$

A) All real numbers

B) $\frac{10}{3}$

C) No solution.

D) $-3\frac{1}{12}$

43) Find the value of x that satisfies the below equation

$$\frac{365}{4} = 6\left(\frac{5}{2}v + \frac{10}{3}\right)$$

A) -1　　B) $\frac{19}{4}$

C) $7\frac{1}{9}$　　D) 2

44) Find the value of x that satisfies the below equation

$$-\frac{21967}{240} = -\frac{11}{3}\left(\frac{39}{8}r + \frac{7}{5}\right)$$

A) $\frac{29}{6}$　　B) $1\frac{2}{3}$

C) $8\frac{2}{3}$　　D) 11

45) Find the value of x that satisfies the below equation

$$\frac{9}{4}n + \frac{9}{5}\left(\frac{13}{2}n - \frac{12}{7}\right) = \frac{3798}{35}$$

A) 0

B) $-\frac{1}{2}$

C) All real numbers

D) 8

46) Find the value of x that satisfies the below equation

$$-\frac{12281}{140} = -\frac{3}{4} - 4\left(\frac{22}{5}n + 1\right)$$

A) $\frac{10}{7}$　　B) $7\frac{1}{16}$

C) $\frac{33}{7}$　　D) $1\frac{8}{15}$

Grade 8

Vol 1 Week 11 Linear equations

47) Find the value of x that satisfies the below equation

$$\frac{13}{4}\left(-5v + \frac{5}{3}\right) = -\frac{4615}{48}$$

A) All real numbers

B) $\frac{25}{4}$

C) $\frac{4}{7}$

D) $3\frac{7}{12}$

48) Find the value of x that satisfies the below equation

$$\frac{13}{6}\left(7b + \frac{1}{5}\right) = \frac{533}{5}$$

A) $1\frac{10}{11}$

B) -10

C) 7

D) $2\frac{8}{15}$

49) Find the value of x that satisfies the below equation

$$\frac{17}{4}\left(\frac{5}{2}b + \frac{1}{6}\right) = \frac{16201}{192}$$

A) $2\frac{2}{9}$

B) $\frac{9}{13}$

C) $\frac{63}{8}$

D) $1\frac{4}{11}$

50) Find the value of x that satisfies the below equation

$$\frac{13369}{140} = -\frac{18}{5}\left(\frac{24}{7}x - \frac{15}{8}\right) + \frac{5}{4}x$$

A) -8

B) $7\frac{4}{7}$

C) $-\frac{5}{3}$

D) All real numbers

51) Find the value of x that satisfies the below equation

$$\frac{36223}{240} = -\frac{7}{2}x + \frac{17}{2}\left(\frac{13}{4}x + \frac{6}{5}\right)$$

A) $\frac{35}{6}$

B) $\frac{3}{2}$

C) -11

D) 3

52) Find the value of x that satisfies the below equation

$$\frac{4071}{40} = \frac{29}{8}n - \frac{14}{5}\left(-\frac{25}{7}n - \frac{16}{7}\right)$$

A) $2\frac{1}{3}$

B) No solution.

C) $\frac{12}{13}$

D) 7

Grade 8

Vol 1 Week 11 Linear equations

53) Find the value of x that satisfies the below equation

$$-\frac{2515}{21} = 5\left(\frac{17}{6}x + \frac{1}{3}\right)$$

A) $6\frac{3}{4}$ B) $-\frac{60}{7}$

C) No solution. D) 0

54) Find the value of x that satisfies the below equation

$$\frac{1261}{15} = -\frac{13}{6}\left(\frac{9}{2}n + 1\right) - \frac{1}{5}n$$

A) No solution. B) $-3\frac{3}{16}$

C) $-\frac{26}{3}$ D) -2

55) Find the value of x that satisfies the below equation

$$-6x - \frac{13}{8}\left(\frac{9}{2}x + \frac{3}{2}\right) = \frac{1665}{16}$$

A) $4\frac{5}{14}$ B) $-1\frac{9}{10}$

C) -8 D) -1

56) Find the value of x that satisfies the below equation

$$\frac{9}{2}\left(7v + \frac{1}{2}\right) + \frac{8}{7}v = -\frac{5421}{28}$$

A) -6

B) $-\frac{2}{7}$

C) $\frac{2}{11}$

D) All real numbers

57) Find the value of x that satisfies the below equation

$$-\frac{1}{8}v + 5\left(7v + \frac{33}{8}\right) = \frac{1725}{16}$$

A) $3\frac{1}{13}$ B) $-\frac{16}{11}$

C) $\frac{5}{2}$ D) -2

58) Find the value of x that satisfies the below equation

$$6\left(\frac{14}{5}x + \frac{23}{6}\right) = \frac{514}{5}$$

A) -1

B) All real numbers

C) $\frac{2}{3}$

D) $\frac{19}{4}$

Grade 8

**Vol 1
Week 11
Linear equations**

59) Find the value of x that satisfies the below equation

$$\frac{1365}{16} = \frac{15}{2}\left(\frac{35}{8}n + \frac{9}{2}\right)$$

A) All real numbers

B) $\frac{11}{7}$

C) $-\frac{8}{7}$

D) $-2\frac{9}{14}$

60) Find the value of x that satisfies the below equation

$$\frac{6293}{72} = \frac{14}{3}\left(\frac{35}{8}x + \frac{1}{2}\right)$$

A) $\frac{25}{6}$

B) $-\frac{2}{5}$

C) $\frac{19}{14}$

D) -2

61) Find the value of x that satisfies the below equation

$$\frac{1101}{10} = -6\left(-n - \frac{13}{5}\right) + 8n$$

A) $4\frac{5}{9}$

B) $6\frac{1}{2}$

C) $1\frac{3}{7}$

D) $\frac{27}{4}$

62) Find the value of x that satisfies the below equation

$$-\frac{469}{3} = \frac{14}{3}\left(\frac{9}{2}r + \frac{5}{2}\right)$$

A) -8

B) $-\frac{12}{7}$

C) -1

D) $8\frac{5}{6}$

63) Find the value of x that satisfies the below equation

$$-\frac{25}{4}\left(\frac{23}{6}m + \frac{1}{2}\right) + \frac{5}{3} = \frac{665}{4}$$

A) All real numbers

B) -7

C) $5\frac{5}{8}$

D) No solution.

64) Find the value of x that satisfies the below equation

$$\frac{11}{3}\left(\frac{22}{5}x + \frac{21}{8}\right) + \frac{4}{3}x = \frac{11111}{120}$$

A) $3\frac{5}{6}$

B) $\frac{19}{4}$

C) $1\frac{1}{6}$

D) $1\frac{1}{2}$

Grade 8

**Vol 1
Week 11
Linear equations**

65) Find the value of x that satisfies the below equation

$$-\frac{569}{6} = -\frac{7}{2}\left(\frac{11}{2}n + 1\right) - \frac{3}{2}$$

A) No solution. B) $-\frac{15}{8}$

C) $1\frac{1}{2}$ D) $\frac{14}{3}$

66) The sum of two consecutive multiples of 3 is 15. Find the two multiples.

67) The sum of two natural numbers is 49. The numbers are in the ratio 3 : 4. Find the numbers.

68) $\frac{1}{4}$ th of a number is 12 less than the original number. Find the number.

69) Find the value of x that satisfies the below equation

$$\frac{31}{8}b + \frac{15}{4}\left(7b + \frac{1}{4}\right) = -\frac{5965}{64}$$

A) $-\frac{25}{8}$ B) $1\frac{4}{5}$

C) $-\frac{11}{13}$ D) -1

70) Three numbers are in the ratio 1 : 2 : 3. If the sum of the largest and the smallest numbers is 24, find these numbers.

71) Find two consecutive even numbers whose sum is equal to 58.

72) The product of two numbers is 192. Their quotient is $\frac{4}{3}$. Find the numbers.

Grade 8

Vol 1 Week 11 Linear equations

73) Brian's father is 28 years older than him. After 10 years Brian is exactly half his father's age. Find the present ages of Brian and his father.

74) Four sevenths of a number is 14 more than half of the number. Find the number.

75) The sum of three numbers is 90 and their ratio is 2:5:8. Find the numbers

76) Two numbers are such that the ratio between them is 8:3. If the sum of the numbers is 143, find the numbers.

77) Three numbers are in the ratio 2:3:4. The sum of their cubes is 792. Find the numbers.

78) Two numbers are in the ratio 2:3. If each number is increased by 6, the ratio of the new numbers so formed is 4:5. Find the original numbers.

79) The sum of three consecutive even numbers is 234. Find the numbers.

80) 150 is to be divided into three parts in such a way that the second number is five - sixths the first and third number is four - fifths the second. Find the numbers.

Grade 8

Vol 1 Week 11 Linear equations

81) The sum of digits of a 2-digit number is 15. If the new number formed by interchanging the digits is less than the original number by 27, find the original number.

82) The numerator of a rational number is less than its denominator by 3. If 3 is subtracted from the numerator and 2 is added to its denominator, the number becomes 1/5. Find the original number.

83) Three numbers are in the ratio 4:5:6. If the sum of the largest and the smallest equals the sum of the third and 55, find the numbers.

84) George's father is four times as old as George. After five years, his father will be three time as old as George will be then. Find their present ages.

85) Two - thirds of a number is 20 less than the original number. Find the number.

86) The distance between two cities is 300 miles. Two cars start simultaneously from these cities on the same road to cross each other. The speed of one of them is greater than that of the other by 5 miles/hr. If the distance between the two cars after 2 hours of their start is 62 miles, find the speed of each car.

Grade 8

**Vol 1
Week 11
Linear equations**

87) The difference between two natural numbers is 34. The ratio of these two numbers is 1:3. Find these numbers.

88) The distance between two cities is 300 miles. Two cars start simultaneously to cross each other. The speed of one of them is greater than that of other by 7 miles/hr. If the distance between the 2 cars after 2 hours of their start is 34 miles, find the speed of each car.

89) The sum of three consecutive even numbers is 36. What is the smallest of these numbers?

   A) 8   B) 12

   C) 14   D) 10

90) The sum of three consecutive odd numbers is 69. What is the smallest of these numbers?

   A) 21   B) 23

   C) 19   D) 25

91) The sum of three consecutive numbers is 51. What is the smallest of these numbers?

   A) 17   B) 16

   C) 18   D) 14

92) The sum of three consecutive odd numbers is 33. What is the smallest of these numbers?

   A) 9   B) 11

   C) 13   D) 7

93) The sum of three consecutive odd numbers is 57. What is the smallest of these numbers?

   A) 15   B) 19

   C) 21   D) 17

94) Jacob had $20 to spend on eight pens. After buying them he had $4. How much did each pen cost?

   A) $2.50   B) $1

   C) $0.50   D) $2

Grade 8

Vol 1 Week 11 Linear equations

95) The sum of three consecutive odd numbers is 39. What is the smallest of these numbers?

   A) 9   B) 13

   C) 11   D) 15

96) The sum of three consecutive numbers is 57. What is the smallest of these numbers?

   A) 18   B) 20

   C) 19   D) 16

97) The sum of three consecutive numbers is 78. What is the smallest of these numbers?

   A) 27   B) 26

   C) 23   D) 25

98) The sum of three consecutive even numbers is 84. What is the smallest of these numbers?

   A) 26   B) 24

   C) 28   D) 30

99) The sum of three consecutive numbers is 30. What is the smallest of these numbers?

   A) 11   B) 10

   C) 7   D) 9

100) The sum of three consecutive even numbers is 30. What is the smallest of these numbers?

   A) 6   B) 12

   C) 10   D) 8

101) The sum of three consecutive numbers is 69. What is the smallest of these numbers?

   A) 23   B) 22

   C) 20   D) 24

102) The sum of three consecutive numbers is 45. What is the smallest of these numbers?

   A) 16   B) 15

   C) 14   D) 12

Grade 8

**Vol 1
Week 11
Linear equations**

103) The sum of three consecutive numbers is 33. What is the smallest of these numbers?

   A) 10   B) 11
   C) 12   D) 8

104) The sum of three consecutive even numbers is 90. What is the smallest of these numbers?

   A) 28   B) 32
   C) 30   D) 26

105) The sum of three consecutive even numbers is 54. What is the smallest of these numbers?

   A) 18   B) 20
   C) 14   D) 16

106) The sum of three consecutive numbers is 66. What is the smallest of these numbers?

   A) 22   B) 23
   C) 19   D) 21

107) The sum of three consecutive odd numbers is 87. What is the smallest of these numbers?

   A) 27   B) 31
   C) 29   D) 25

108) 378 students went on a field trip. Nine buses were filled and 18 students traveled in cars. How many students were in each bus?

   A) 40   B) 33
   C) 42   D) 36

109) Tom said, "400 reduced by 3 times my age is 124." What is his age?

   A) 113   B) 28
   C) 104   D) 92

110) Cathy said, 400 reduced by 2 times my age is 252. How old am I?

   A) 74   B) 81
   C) 87   D) 104

111) Lola said, 400 reduced by 3 times my age is 184. How old am I?

   A) 152   B) 72
   C) 70   D) 76

112) Ron said, "300 reduced by 4 times my age is 80." What is his age?

   A) 50   B) 57
   C) 20   D) 55

©All rights reserved-Math-Knots LLC., VA-USA

www.math-knots.com | www.a4ace.com

Grade 8

Vol 1
Week 11
Linear equations

113) Wilson said, "400 reduced by 4 times my age is 64." What is his age?

A) 84   B) 91

C) 103   D) 144

114) Mary said, 500 reduced by 3 times my age is 341. How old am I?

A) 63   B) 53

C) 523   D) 56

115) Zara said, 400 reduced by 2 times my age is 274. How old am I?

A) 148   B) 64

C) 59   D) 63

116) Rick said, "500 reduced by 4 times my age is 204." What is his age?

A) 316   B) 84

C) 66   D) 74

117) Chris said, 400 reduced by 4 times my age is 24. How old am I?

A) 107   B) 116

C) 304   D) 94

118) Bella said, 400 reduced by 3 times my age is 208. How old am I?

A) 224   B) 67

C) 70   D) 64

119) The sum of three consecutive even numbers is 60. What is the smallest of these numbers?

A) 18   B) 22

C) 20   D) 16

120) You had $21 to spend on four pencils. After buying them you had $5. How much did each pencil cost?

A) $5   B) $4

C) $1.25   D) $5.25

121) The sum of three consecutive odd numbers is 93. What is the smallest of these numbers?

A) 33   B) 29

C) 27   D) 31

122) You had $22 to spend on two raffle tickets. After buying them you had $16. How much did each raffle ticket cost?

A) $3   B) $8

C) $11   D) $2

Grade 8

Vol 1 Week 12 Coordinates

1) Find the distance between the points P(-4,11) and Q(-3,4)

   A) $4\sqrt{11}$ B) $5\sqrt{2}$
   C) $2\sqrt{2}$ D) $\sqrt{22}$

2) Find the distance between the points P(0,-7) and Q(-3,-3)

   A) $6\sqrt{5}$ B) $\sqrt{7}$
   C) 5   D) $\sqrt{109}$

3) Find the distance between the points P(11,-10) and Q(-3,6)

   A) $\sqrt{30}$ B) $\sqrt{13}$
   C) $2\sqrt{113}$ D) $4\sqrt{3}$

4) Find the distance between the points P(-10,0) and Q(-2,4)

   A) $2\sqrt{3}$ B) 4
   C) $4\sqrt{5}$ D) $8\sqrt{2}$

5) Find the distance between the points P(-8,-7) and Q(10,-11)

   A) $\sqrt{22}$ B) $8\sqrt{5}$
   C) $2\sqrt{82}$ D) $2\sqrt{85}$

6) Find the distance between the points P(-3,3) and Q(-2,10)

   A) $3\sqrt{2}$ B) 12
   C) $2\sqrt{2}$ D) $5\sqrt{2}$

7) Find the distance between the points P(1,-7) and Q(0,0)

   A) $5\sqrt{2}$ B) $\sqrt{37}$
   C) $2\sqrt{2}$ D) $\sqrt{65}$

8) Find the distance between the points P(3,-1) and Q(-5,9)

   A) $22\sqrt{2}$ B) $3\sqrt{2}$
   C) $2\sqrt{17}$ D) $2\sqrt{41}$

9) Find the distance between the points P(-10,9) and Q(6,-7)

   A) 17   B) $2\sqrt{5}$
   C) $4\sqrt{2}$ D) $16\sqrt{2}$

10) Find the distance between the points P(-1,8) and Q(-5,3)

   A) 3   B) $\sqrt{17}$
   C) $\sqrt{41}$ D) $2\sqrt{41}$

©All rights reserved-Math-Knots LLC., VA-USA

Grade 8

**Vol 1
Week 12
Coordinates**

11) Find the distance between the points P(1,7) and Q(-9,9)

   A) $2\sqrt{6}$ B) $2\sqrt{26}$

   C) $8\sqrt{5}$ D) $2\sqrt{3}$

12) Find the distance between the points P(11,5) and Q(11,10)

   A) 5   B) $\sqrt{7}$

   C) $\sqrt{5}$ D) $3\sqrt{2}$

13) Find the distance between the points P(-11,-9) and Q(7,-10)

   A) $5\sqrt{13}$ B) $4\sqrt{2}$

   C) $4\sqrt{10}$ D) $\sqrt{19}$

14) Find the distance between the points P(-2,2) and Q(1,10)

   A) $\sqrt{26}$ B) $\sqrt{13}$

   C) $\sqrt{11}$ D) $\sqrt{73}$

15) Find the distance between the points P(5,11) and Q(1,4)

   A) $3\sqrt{29}$ B) $\sqrt{11}$

   C) 3   D) $\sqrt{65}$

16) Find the distance between the points P(-3,-6) and Q(-11,-11)

   A) $\sqrt{93}$ B) $\sqrt{89}$

   C) $\sqrt{13}$ D) $\sqrt{3}$

17) Find the distance between the points P(6,3) and Q(7,9)

   A) $\sqrt{37}$ B) $\sqrt{5}$

   C) 5   D) $\sqrt{7}$

18) Find the distance between the points P(0,3) and Q(-2,11)

   A) $10\sqrt{2}$ B) 4

   C) $2\sqrt{17}$ D) $\sqrt{10}$

19) Find the distance between the points P(8,-4) and Q(10,-8)

   A) $\sqrt{6}$ B) $\sqrt{30}$

   C) $6\sqrt{5}$ D) $2\sqrt{5}$

20) Find the distance between the points P(7.2,-1.7) and Q(-0.2,-2). (Round to the nearest tenth)

   A) 3.3   B) 2.8

   C) 7.4   D) 5.9

Grade 8

Vol 1 Week 12 Coordinates

21) Find the distance between the points P(8,3) and Q(11,-5)

   A) $\sqrt{21}$ B) $\sqrt{73}$

   C) $\sqrt{11}$ D) $2\sqrt{10}$

22) Find the distance between the points P(12,-8) and Q(4,-2)

   A) $\sqrt{14}$ B) 10

   C) $2\sqrt{89}$ D) $\sqrt{26}$

23) Find the distance between the points P(-1,1) and Q(-5,8)

   A) $\sqrt{15}$ B) $3\sqrt{13}$

   C) $\sqrt{11}$ D) $\sqrt{65}$

24) Find the distance between the points P(7.4,7.5) and Q(-1.5,0.7). (Round to the nearest tenth)

   A) 4   B) 10.1

   C) 1.5   D) 11.2

25) Find the distance between the points P(-4.05,2.8) and Q(-1.3,0.25). (Round to the nearest tenth)

   A) 2.9   B) 6.2

   C) 3.8   D) 2.3

26) Find the distance between the points P(-3.2,-6) and Q(4.2,5.2). (Round to the nearest tenth)

   A) 1.3   B) 4.3

   C) 13.4   D) 1.9

27) Find the distance between the points P(-1.5,6.1) and Q(1.5,2.6). (Round to the nearest tenth)

   A) 2.5   B) 4.6

   C) 2.9   D) 8.7

28) Find the distance between the points P(1,-1.96) and Q(3.9,-2.9). (Round to the nearest tenth)

   A) 3   B) 6.9

   C) 2   D) 3.1

29) Find the distance between the points P(2.3074,-2.8) and Q(0.1,1.7). (Round to the nearest tenth)

   A) 5   B) 8.8

   C) 10.2   D) 2.6

30) Find the distance between the points P(-1.654,-2.6) and Q(-2.5,-6.6). (Round to the nearest tenth)

   A) 4.1   B) 8.2

   C) 2.2   D) 10.1

©All rights reserved-Math-Knots LLC., VA-USA   www.math-knots.com | www.a4ace.com

Grade 8

**Vol 1
Week 12
Coordinates**

31) Find the distance between the points P(-0.42,-5.1) and Q(1.9,6.85).
(Round to the nearest tenth)

    A) 12.2   B) 3.8

    C) 3.1   D) 2.3

32) Find the distance between the points P(2,-0.96) and Q(-4.8,-4.326).
(Round to the nearest tenth)

    A) 7.6   B) 6

    C) 1.9   D) 3.2

33) Find the distance between the points P(4.5,-7.4) and Q(-6.29,0.16).
(Round to the nearest tenth)

    A) 7   B) 9.2

    C) 13.2   D) 4.3

34) Find the distance between the points P(-7.9,3.661) and Q(7.8,2.2).
(Round to the nearest tenth)

    A) 15.8   B) 5.9

    C) 12.4   D) 4.1

35) Find the distance between the points plotted below

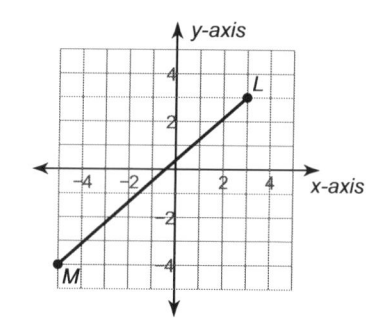

    A) $\sqrt{3}$ B) $\sqrt{113}$

    C) $\sqrt{15}$ D) 1

36) Find the distance between the points plotted below

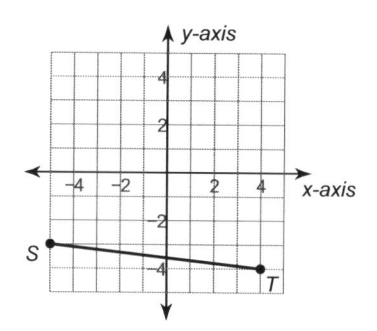

    A) $\sqrt{10}$ B) $4\sqrt{3}$

    C) $\sqrt{73}$ D) $\sqrt{82}$

37) Find the distance between the points plotted below

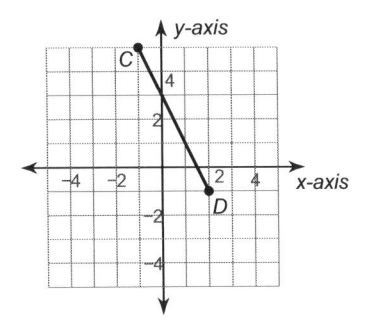

A) 3   B) $\sqrt{17}$

C) $3\sqrt{5}$ D) $\sqrt{15}$

39) Find the distance between the points plotted below

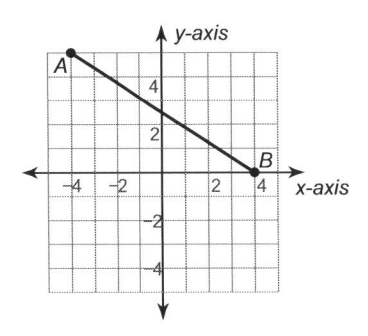

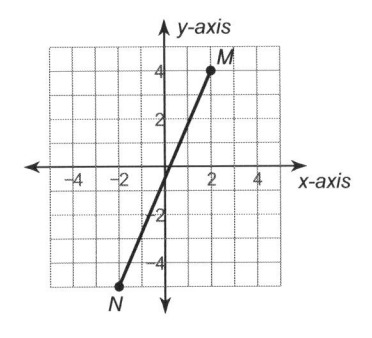

A) 5   B) $\sqrt{13}$

C) $3\sqrt{10}$ D) $\sqrt{89}$

38) Find the distance between the points plotted below

A) 1   B) $\sqrt{13}$

C) $\sqrt{5}$ D) $\sqrt{97}$

40) Find the distance between the points plotted below

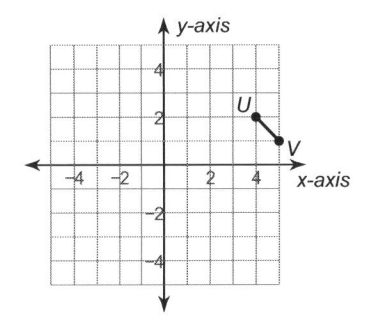

A) $\sqrt{6}$ B) $6\sqrt{2}$

C) 10   D) $\sqrt{2}$

41) Find the distance between the points plotted below

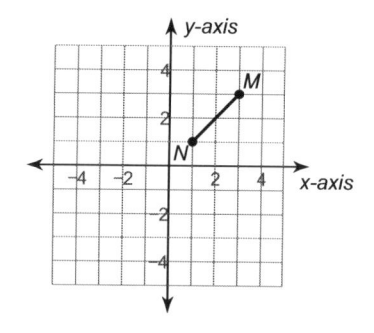

A) 2   B) 2√2

C) 10   D) 4√2

43) Find the distance between the points plotted below

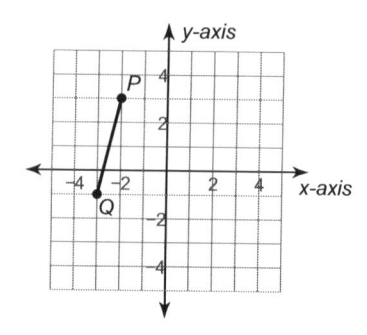

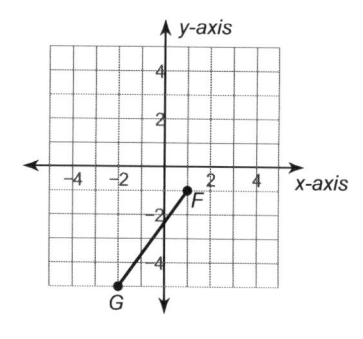

A) √29   B) √7

C) √5   D) √17

42) Find the distance between the points plotted below

A) √37   B) √7

C) 5   D) √5

44) Find the distance between the points plotted below

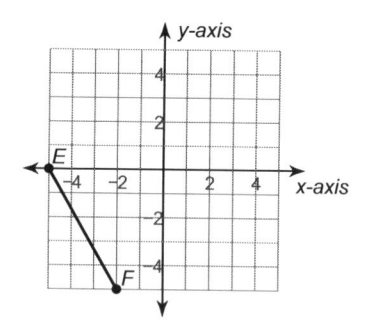

A) 2√6   B) 2√2

C) √34   D) √74

Grade 8

Vol 1
Week 12
Coordinates

45) Find the distance between the points plotted below

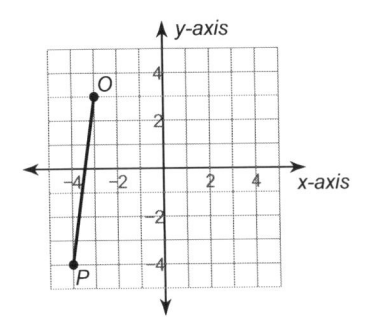

A) $5\sqrt{2}$ B) $2\sqrt{2}$

C) $4\sqrt{3}$ D) $\sqrt{6}$

46) Find the distance between the points plotted below

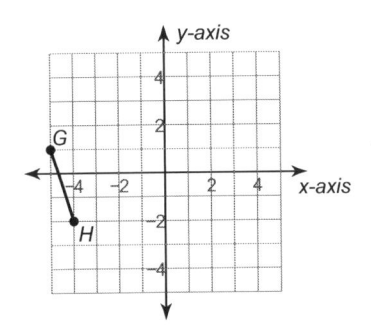

A) $\sqrt{82}$ B) 2

C) $\sqrt{10}$ D) $2\sqrt{2}$

47) Find the distance between the points plotted below

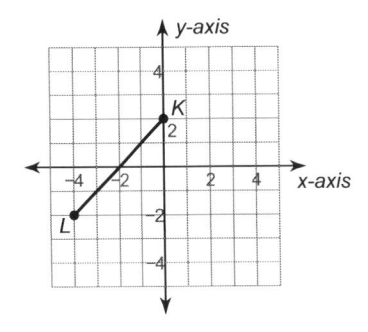

A) 4   B) 2

C) $4\sqrt{2}$ D) $2\sqrt{2}$

48) Find the distance between the points plotted below

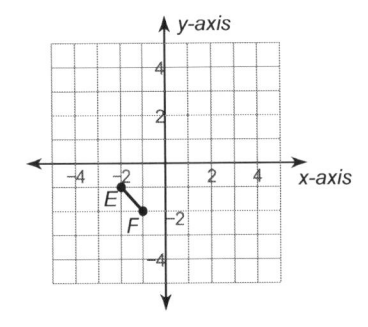

A) $3\sqrt{2}$ B) $\sqrt{2}$

C) $2\sqrt{2}$ D) $\sqrt{58}$

©All rights reserved-Math-Knots LLC., VA-USA   www.math-knots.com | www.a4ace.com

Grade 8

Vol 1 — Week 12 — Coordinates

49) Find the distance between the points plotted below

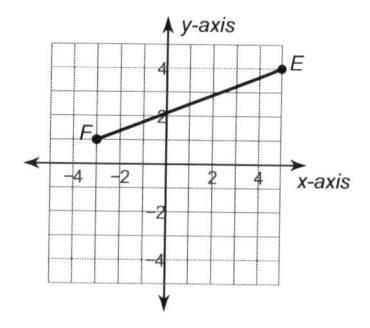

A) $\sqrt{11}$ B) $\sqrt{21}$

C) $\sqrt{5}$ D) $\sqrt{73}$

50) Find the distance between the points plotted below

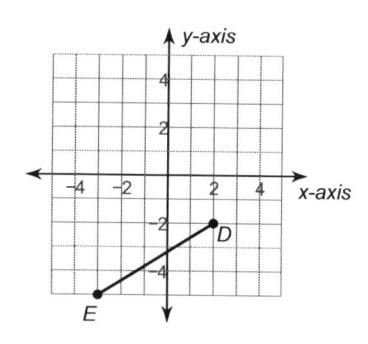

A) $\sqrt{6}$ B) $5\sqrt{2}$

C) $\sqrt{34}$ D) $2\sqrt{2}$

51) Find the distance between the points plotted below

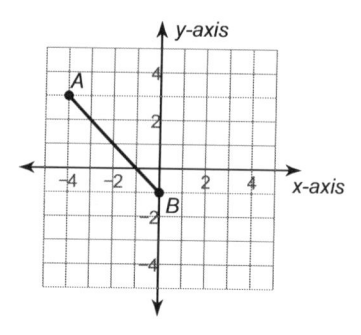

A) $2\sqrt{3}$ B) $2\sqrt{5}$

C) $4\sqrt{2}$ D) $2\sqrt{2}$

52) Find the distance between the points plotted below

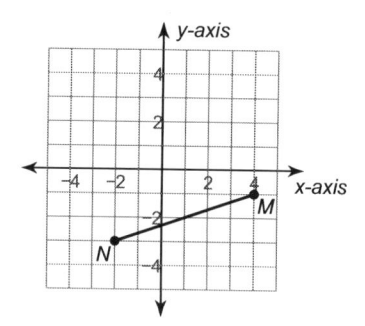

A) $2\sqrt{2}$ B) $\sqrt{85}$

C) $2\sqrt{10}$ D) $\sqrt{6}$

Grade 8

Vol 1
Week 12
Coordinates

53) Find the distance between the points plotted below

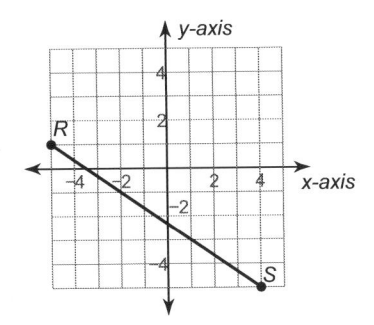

A) $\sqrt{17}$ B) $3\sqrt{13}$

C) $\sqrt{15}$ D) $3\sqrt{2}$

54) Find the distance between the points plotted below

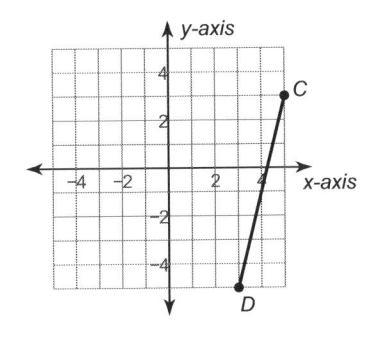

A) $2\sqrt{17}$ B) 5

C) $\sqrt{5}$ D) $\sqrt{10}$

55) Find the distance between the points plotted below

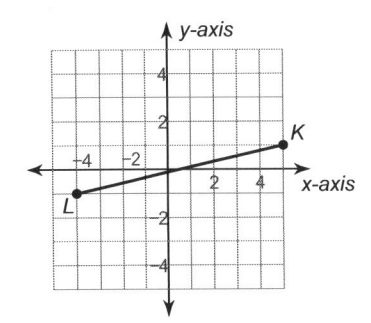

A) 1   B) $\sqrt{11}$

C) $\sqrt{7}$ D) $\sqrt{85}$

56) Find the distance between the points plotted below

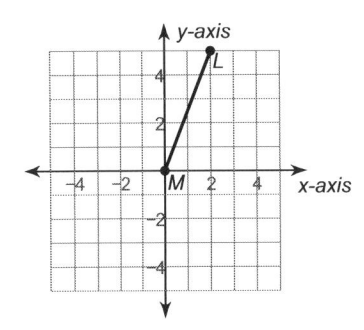

A) $\sqrt{21}$ B) $\sqrt{3}$

C) $\sqrt{7}$ D) $\sqrt{29}$

Grade 8

Vol 1
Week 12
Coordinates

57) Find the distance between the points plotted below

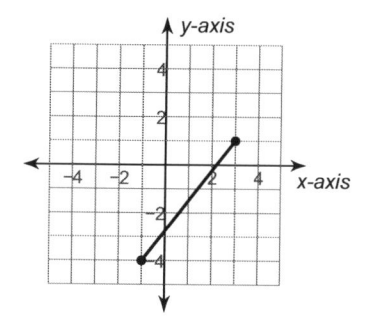

A) 3 B) √13

C) √5 D) √41

58) Find the distance between the points plotted below

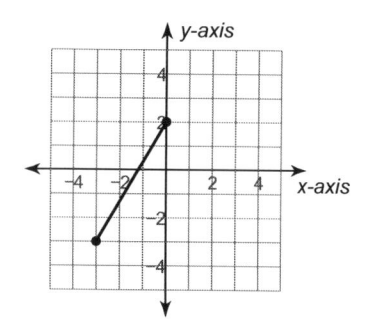

A) √34 B) √2

C) √10 D) 2√2

59) Find the distance between the points plotted below

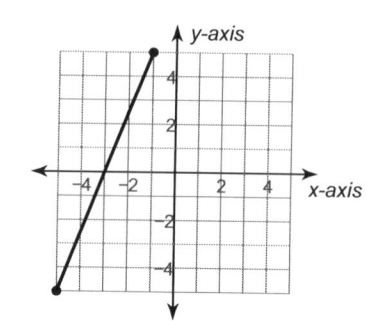

A) √6 B) 6

C) 2√29 D) √14

60) Find the distance between the points plotted below

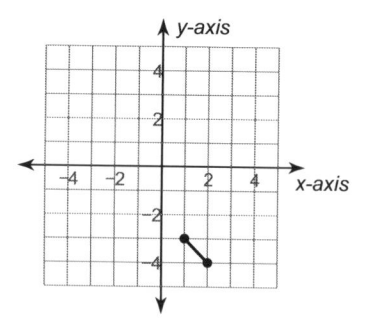

A) √2 B) √58

C) √10 D) 2√10

Grade 8

Vol 1 Week 12 Coordinates

61) Find the distance between the points plotted below

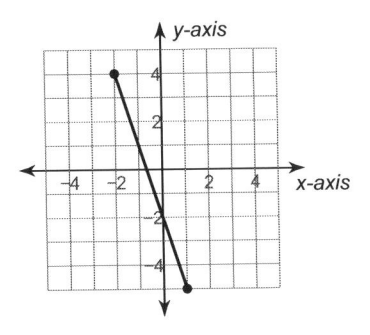

A) $4\sqrt{2}$ B) $3\sqrt{10}$

C) $2\sqrt{3}$ D) $\sqrt{2}$

63) Find the distance between the points plotted below

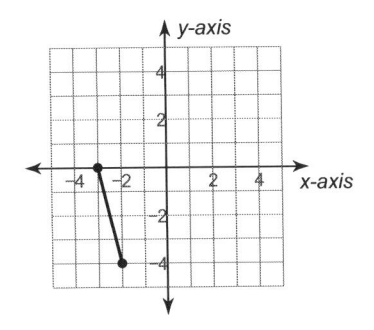

A) $\sqrt{41}$ B) 3

C) $\sqrt{17}$ D) $\sqrt{5}$

62) Find the distance between the points plotted below

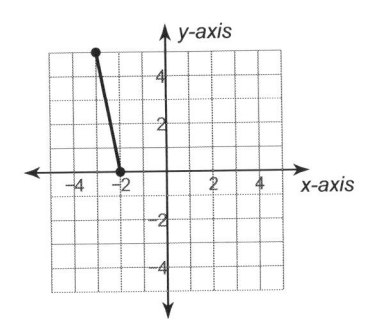

A) $\sqrt{10}$ B) $\sqrt{26}$

C) $\sqrt{6}$ D) $5\sqrt{2}$

64) Find the distance between the points plotted below

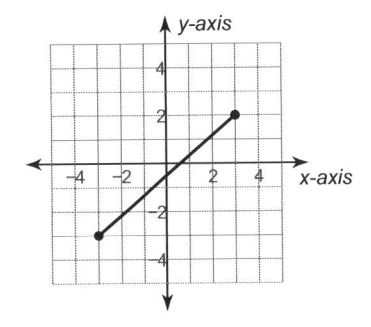

A) $\sqrt{61}$ B) 1

C) $5\sqrt{2}$ D) $\sqrt{11}$

©All rights reserved-Math-Knots LLC., VA-USA

Grade 8

Vol 1 Week 12 Coordinates

65) Find the midpoint of the line segment plotted below

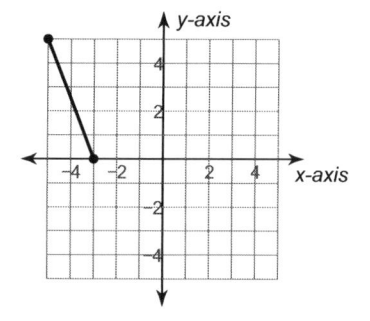

67) Find the midpoint of the line segment plotted below

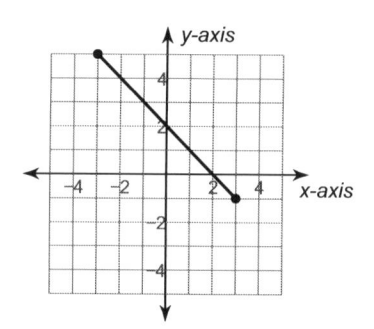

66) Find the midpoint of the line segment plotted below

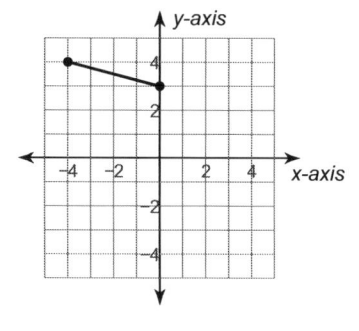

68) Find the midpoint of the line segment plotted below

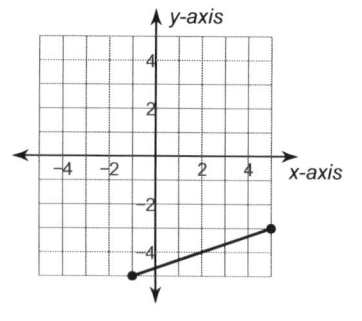

Grade 8

**Vol 1
Week 12
Coordinates**

69) Find the midpoint of the line segment plotted below

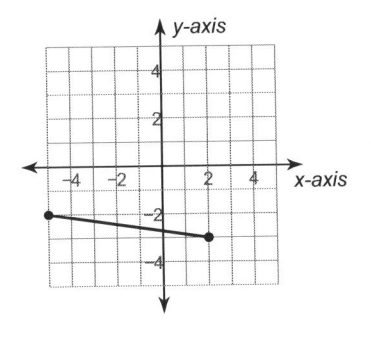

71) Find the midpoint of the line segment plotted below

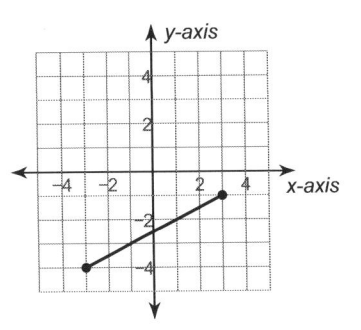

70) Find the midpoint of the line segment plotted below

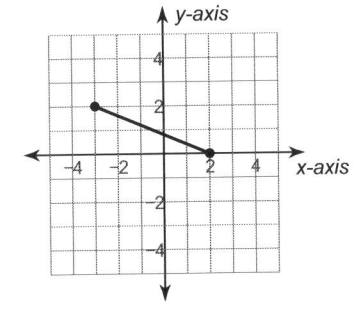

72) Find the midpoint of the line segment plotted below

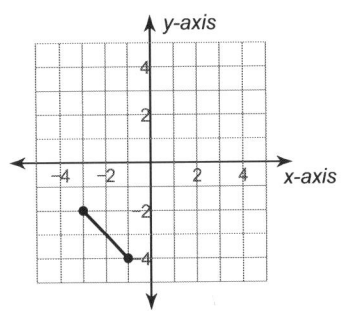

Grade 8

Vol 1 Week 12
Coordinates

73) Find the midpoint of the line segment plotted below

77) Find the midpoint of the line segment plotted below

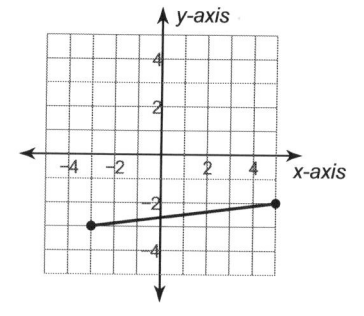

74) Find the midpoint of the points P(11,14) and Q(-7,2)

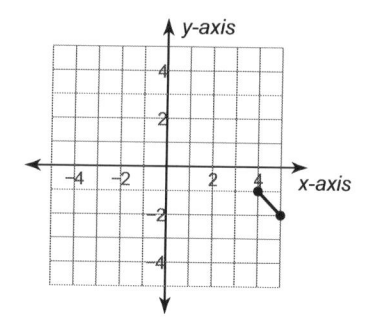

78) Find the midpoint of the points P(-3,11) and Q(3,12)

75) Find the midpoint of the points P(14,8) and Q(-6,4)

79) Find the midpoint of the points P(3,-6) and Q(-9,0)

76) Find the midpoint of the points P(9,-3) and Q(14,7)

80) Find the midpoint of the points P(-14,3) and Q(-2,5)

Grade 8

Vol 1　Week 12　Coordinates

81) Find the midpoint of the points P(7,12) and Q(14,-14)

82) Find the midpoint of the points P(5,13) and Q(7,1)

83) Find the midpoint of the points P(-8,0) and Q(-14,-14)

84) Find the midpoint of the points P(0,10) and Q(14,-13)

85) Find the midpoint of the points P(-8,8) and Q(-10,-14)

86) Find the midpoint of the points P(8,2) and Q(4,7)

87) Find the midpoint of the points P(-10,12) and Q(-8,4)

88) Find the midpoint of the points P(14,20) and Q(8,-2)

89) Find the midpoint of the points P(20,-12) and Q(14,-4)

90) Find the midpoint of the points P(-2,3) and Q(-6,-9)

Grade 8

Vol 1 | Week 12 — Coordinates

91) Find the midpoint of the points P(3,-7) and Q(10,20)

92) Find the midpoint of the points P(7,18) and Q(8,2)

93) Find the midpoint of the points P(-2,-18) and Q(0,-8)

94) Find the midpoint of the points P(19,-1) and Q(11,-17)

95) Find the midpoint of the points P(20,-14) and Q(17,8)

96) Find the midpoint of the points P(11,5) and Q(16,-17)

97) Find the midpoint of the points P(-3,17) and Q(10,-15)

98) Find the midpoint of the points P(-13,3) and Q(-15,-18)

99) Find the midpoint of the points P(-9,-3) and Q(16,-19)

100) Find the midpoint of the points P(-16,-3) and Q(-11,10)

Grade 8

Vol 1 Week 12 Coordinates

101) Find the endpoint of the line segment given one endpoint as (20,6) and midpoint as Q(-29,-25).

A) $(-4, 32)$ B) $(69, 37)$

C) $(-78, -56)$ D) $(-4.5, -9.5)$

102) Find the endpoint of the line segment given one endpoint as (35,20) and midpoint as Q(6,-29).

A) $(20.5, -4.5)$ B) $(64, 69)$

C) $(-23, -78)$ D) $(24, -24)$

103) Find the endpoint of the line segment given one endpoint as (-35,33) and midpoint as Q(6,-40).

A) $(-14.5, -3.5)$ B) $(47, -113)$

C) $(-76, 106)$ D) $(-10, 12)$

104) Find the endpoint of the line segment given one endpoint as (13,-38) and midpoint as Q(18,-30).

A) $(23, -22)$ B) $(8, -46)$

C) $(-34, -32)$ D) $(15.5, -34)$

105) Find the endpoint of the line segment given one endpoint as (-38,-36) and midpoint as Q(28,-34).

A) $(-104, -38)$ B) $(-5, -35)$

C) $(34, 6)$ D) $(94, -32)$

106) Find the endpoint of the line segment given one endpoint as (-30,13) and midpoint as Q(17,-35).

A) $(-77, 61)$ B) $(-40, -32)$

C) $(64, -83)$ D) $(-6.5, -11)$

107) Find the endpoint of the line segment given one endpoint as (4,27) and midpoint as Q(2,-21).

A) $(6, 75)$ B) $(0, -69)$

C) $(-37, -27)$ D) $(3, 3)$

108) Find the endpoint of the line segment given one endpoint as (12,-16) and midpoint as Q(-34,-15).

A) $(58, -17)$ B) $(-11, -15.5)$

C) $(6, 20)$ D) $(-80, -14)$

Grade 8

Vol 1, Week 13 — System of equations

1) Solve the below system of equations by using the method of elimination

$$-5x + 6y = -14$$
$$3x - 6y = 6$$

A) $(4, 1)$

B) Infinite number of solutions

C) $(-4, 1)$

D) $(1, -4)$

2) Solve the below system of equations by using the method of elimination

$$4x - 6y = -4$$
$$-4x + 2y = 12$$

A) $(4, -2)$ B) $(-4, 2)$

C) $(-4, -2)$ D) $(-2, -4)$

3) Solve the below system of equations by using the method of elimination

$$7x - 4y = 16$$
$$-7x + y = -25$$

A) $(-6, -4)$ B) $(4, 3)$

C) $(-4, -6)$ D) $(4, -6)$

4) Solve the below system of equations by using the method of elimination

$$x - 3y = 2$$
$$-x + 9y = -8$$

A) $(-7, 1)$ B) $(7, -1)$

C) $(-1, -1)$ D) $(-1, -7)$

5) Solve the below system of equations by using the method of elimination

$$2x - 7y = -3$$
$$-2x - 8y = -12$$

A) $(-2, -1)$ B) $(2, -1)$

C) $(2, 1)$ D) $(-1, 2)$

6) Solve the below system of equations by using the method of elimination

$$5x - 2y = -16$$
$$-5x - 3y = 26$$

A) $(-4, -2)$ B) $(-4, 10)$

C) $(-4, -10)$ D) $(10, 4)$

7) Solve the below system of equations by using the method of elimination

$$x - 2y = -2$$
$$-x - 4y = -16$$

A) $(4, 3)$ B) $(-3, -9)$

C) $(-9, 3)$ D) $(5, -9)$

8) Solve the below system of equations by using the method of elimination

$$8x - 8y = -25$$
$$-8x + 8y = 16$$

A) $(4, -4)$

B) $(4, 4)$

C) No solution

D) Infinite number of solutions

Grade 8

**Vol 1
Week 13
System of equations**

9) Solve the below system of equations by using the method of elimination

$$-10x - 8y = -8$$
$$9x + 8y = 4$$

A) $(-4, -4)$ B) $(-4, 4)$
C) $(4, 4)$ D) $(4, -4)$

10) Solve the below system of equations by using the method of elimination

$$9x + 3y = 3$$
$$-9x - 5y = -11$$

A) $(1, 4)$ B) $(-1, 4)$
C) $(-4, 1)$ D) $(1, -4)$

11) Solve the below system of equations by using the method of elimination

$$-3x + 7y = 28$$
$$3x - y = -4$$

A) $(10, 0)$ B) $(0, 4)$
C) $(0, 10)$ D) $(0, -10)$

12) Solve the below system of equations by using the method of elimination

$$-3x + 3y = 5$$
$$3x - 3y = -9$$

A) $(1, 9)$

B) $(1, -9)$

C) No solution

D) Infinite number of solutions

13) Solve the below system of equations by using the method of elimination

$$2x + 6y = -6$$
$$-2x - 9y = 0$$

A) $(-6, 2)$

B) $(9, 2)$

C) $(-9, 2)$

D) Infinite number of solutions

14) Solve the below system of equations by using the method of elimination

$$-6x + 10y = 10$$
$$6x - y = 26$$

A) $(5, -4)$ B) $(5, 4)$
C) $(-3, 4)$ D) $(-5, 4)$

15) Solve the below system of equations by using the method of elimination

$$-x + 7y = -23$$
$$-2x - 7y = -4$$

A) $(3, -2)$ B) $(-9, 2)$
C) $(9, -2)$ D) $(9, 2)$

16) Solve the below system of equations by using the method of elimination

$$2x - 4y = -9$$
$$-2x + 4y = 18$$

A) $(6, -9)$ B) $(4, -9)$
C) No solution D) $(6, 4)$

Grade 8

Vol 1 Week 13 System of equations

17) Solve the below system of equations by using the method of elimination

$$-2x + 4y = 0$$
$$2x - 10y = 6$$

A) $(2, -1)$ B) $(9, -1)$

C) $(9, 1)$ D) $(-2, -1)$

18) Solve the below system of equations by using the method of elimination

$$-3x - 9y = -21$$
$$3x - 3y = -3$$

A) $(1, 2)$ B) $(2, 1)$

C) $(-1, 9)$ D) No solution

19) Solve the below system of equations by using the method of elimination

$$x - 9y = 4$$
$$-x - 9y = -4$$

A) $(0, 4)$ B) $(4, 0)$

C) $(1, 10)$ D) $(0, -4)$

20) Solve the below system of equations by using the method of elimination

$$-6x + 3y = -6$$
$$6x - y = 6$$

A) $(1, -6)$ B) $(-4, -6)$

C) $(1, 0)$ D) $(4, -6)$

21) Solve the below system of equations by using the method of elimination

$$2x - 8y = 2$$
$$-x + 8y = -9$$

A) $(-7, -2)$ B) $(-7, 2)$

C) $(-2, -7)$ D) $(-2, 7)$

22) Solve the below system of equations by using the method of elimination

$$x + 3y = 1$$
$$-x + y = -13$$

A) $(10, -3)$ B) $(-3, 10)$

C) $(3, -9)$ D) $(3, 10)$

23) Solve the below system of equations by using the method of elimination

$$-4x - 7y = 28$$
$$4x + 7y = -28$$

A) $(-4, 10)$

B) $(10, -4)$

C) $(-4, 8)$

D) Infinite number of solutions

24) Solve the below system of equations by using the method of elimination

$$-2x + 4y = 0$$
$$2x - 6y = -8$$

A) $(8, 4)$

B) Infinite number of solutions

C) $(-4, 7)$

D) $(4, 8)$

Grade 8

Vol 1 Week 13 — System of equations

25) Solve the below system of equations by using the method of elimination

$$-x - 5y = -14$$
$$x + 7y = 18$$

A) $(-4, 2)$ B) $(4, 2)$
C) $(-9, 2)$ D) $(1, 2)$

26) Solve the below system of equations by using the method of elimination

$$7x - 5y = 12$$
$$7x - 5y = 3$$

A) No solution
B) Infinite number of solutions
C) $(4, -9)$
D) $(-4, -9)$

27) Solve the below system of equations by using the method of elimination

$$3x - 6y = 30$$
$$3x - 6y = 30$$

A) $(-1, -9)$
B) Infinite number of solutions
C) $(1, -9)$
D) No solution

28) Solve the below system of equations by using the method of elimination

$$-x - 5y = -25$$
$$5x - 5y = 5$$

A) $(-4, 5)$ B) $(4, 5)$
C) $(-8, 4)$ D) $(5, 4)$

29) Solve the below system of equations by using the method of elimination

$$3x + 8y = -29$$
$$3x - 6y = 27$$

A) $(1, -4)$ B) $(-4, 1)$
C) $(4, -2)$ D) $(-2, -4)$

30) Solve the below system of equations by using the method of elimination

$$-8x + y = 13$$
$$5x + y = -13$$

A) $(-2, 3)$ B) $(3, 2)$
C) $(3, -2)$ D) $(-2, -3)$

31) Solve the below system of equations by using the method of elimination

$$4x + y = 1$$
$$7x + y = -2$$

A) $(-1, 5)$ B) $(1, 5)$
C) $(-5, 1)$ D) $(5, 1)$

32) Solve the below system of equations by using the method of elimination

$$-2x - 6y = 12$$
$$-2x - 10y = 24$$

A) $(3, -3)$ B) $(-3, 3)$
C) $(9, 3)$ D) $(9, -3)$

Grade 8

Vol 1 — Week 13 — System of equations

33) Solve the below system of equations by using the method of elimination

$$-4x - 8y = -16$$
$$-4x - 3y = -11$$

A) $(-1, -2)$ B) $(1, -2)$
C) $(2, 1)$ D) $(-2, 1)$

34) Solve the below system of equations by using the method of elimination

$$5x + y = 2$$
$$3x + y = 2$$

A) $(2, 6)$ B) $(0, 2)$
C) $(2, 0)$ D) $(2, -6)$

35) Solve the below system of equations by using the method of elimination

$$x + 6y = 17$$
$$-8x + 6y = -28$$

A) $(2, 5)$ B) $(5, 2)$
C) $(5, -2)$ D) $(-2, 5)$

36) Solve the below system of equations by using the method of elimination

$$-10x - 6y = 20$$
$$-4x - 6y = -10$$

A) $(-5, -1)$ B) $(-5, 5)$
C) $(5, -5)$ D) $(-5, -5)$

37) Solve the below system of equations by using the method of elimination

$$x + 5y = 17$$
$$x + 6y = 22$$

A) $(-10, 8)$ B) $(-8, 5)$
C) $(5, -8)$ D) $(-8, 10)$

38) Solve the below system of equations by using the method of elimination

$$-x - 9y = -25$$
$$-x - 9y = -25$$

A) $(5, 9)$
B) $(-9, 5)$
C) Infinite number of solutions
D) $(-1, 5)$

39) Solve the below system of equations by using the method of elimination

$$10x - 7y = 2$$
$$10x - 7y = 2$$

A) No solution
B) Infinite number of solutions
C) $(5, -7)$
D) $(7, 5)$

40) Solve the below system of equations by using the method of elimination

$$-3x - y = -21$$
$$2x - y = 29$$

A) $(-1, 9)$ B) $(10, -9)$
C) $(1, -9)$ D) $(-9, -1)$

Grade 8

Vol 1 Week 13 — System of equations

41) Solve the below system of equations by using the method of elimination

$$-2x - y = 1$$
$$4x - y = 13$$

A) $(-2, 5)$ B) $(-2, -5)$

C) $(10, 5)$ D) $(2, -5)$

42) Solve the below system of equations by using the method of elimination

$$3x + 10y = 7$$
$$3x - 4y = -7$$

A) $(-1, 1)$ B) $(1, -3)$

C) $(-8, -1)$ D) $(1, -1)$

43) Solve the below system of equations by using the method of elimination

$$4x - 2y = -16$$
$$-x - 2y = -11$$

A) $(6, -1)$ B) $(2, -3)$

C) $(2, -1)$ D) $(-1, 6)$

44) Solve the below system of equations by using the method of elimination

$$-10x - 6y = 12$$
$$-x - 6y = -15$$

A) $(5, 3)$ B) $(-3, 3)$

C) $(5, -3)$ D) $(-5, 3)$

45) Solve the below system of equations by using the method of elimination

$$-9x - y = -16$$
$$-9x - y = -26$$

A) $(1, 8)$

B) No solution

C) $(-1, 8)$

D) Infinite number of solutions

46) Solve the below system of equations by using the method of elimination

$$-x - 5y = 13$$
$$-8x - 5y = -1$$

A) $(2, -7)$ B) $(2, -3)$

C) $(2, 3)$ D) $(7, 2)$

47) Solve the below system of equations by using the method of elimination

$$8x + 3y = 19$$
$$x + 3y = -16$$

A) $(3, -7)$ B) $(5, -7)$

C) $(-7, 3)$ D) $(-9, 3)$

48) Solve the below system of equations by using the method of elimination

$$5x + 10y = -30$$
$$5x - 6y = -30$$

A) $(6, -1)$ B) $(6, 0)$

C) $(-6, 0)$ D) $(-3, 6)$

Grade 8

Vol 1, Week 13 — System of equations

49) Solve the below system of equations by using the method of elimination

$$10x + y = -12$$
$$10x + y = -12$$

A) $(1, -9)$

B) $(1, 9)$

C) Infinite number of solutions

D) $(1, -8)$

50) Solve the below system of equations by using the method of elimination

$$-x + 7y = 8$$
$$4x + 14y = 10$$

A) $(1, 1)$ B) $(-1, -3)$

C) $(-1, 7)$ D) $(-1, 1)$

51) Solve the below system of equations by using the method of elimination

$$22x - 4y = -10$$
$$-11x - 8y = 35$$

A) $(3, -1)$ B) $(1, -3)$

C) $(-1, -3)$ D) $(-1, 3)$

52) Solve the below system of equations by using the method of elimination

$$5x - 11y = 1$$
$$x + 2y = 17$$

A) $(-9, 1)$ B) $(9, 4)$

C) $(1, -9)$ D) $(-9, -1)$

53) Solve the below system of equations by using the method of elimination

$$-2x - y = 8$$
$$-2x - 6y = -2$$

A) $(-5, -1)$ B) $(5, 2)$

C) $(5, -2)$ D) $(-5, 2)$

54) Solve the below system of equations by using the method of elimination

$$-22x + 10y = -26$$
$$-11x + 5y = -13$$

A) Infinite number of solutions

B) $(-9, -8)$

C) $(-9, 8)$

D) $(9, 8)$

55) Solve the below system of equations by using the method of elimination

$$-18x - 6y = 36$$
$$9x - 3y = -18$$

A) $(1, 0)$ B) $(-3, 1)$

C) $(-2, 0)$ D) $(-1, 0)$

56) Solve the below system of equations by using the method of elimination

$$-18x + 9y = -18$$
$$6x + 8y = 28$$

A) Infinite number of solutions

B) $(2, 2)$

C) $(-2, 2)$

D) $(2, -2)$

Grade 8

Vol 1 Week 13 System of equations

57) Solve the below system of equations by using the method of elimination

$16x - 11y = 19$
$8x - 8y = 32$

A) Infinite number of solutions

B) $(-5, -9)$

C) $(5, -9)$

D) $(-5, 9)$

58) Solve the below system of equations by using the method of elimination

$-5x - 11y = -11$
$-10x + 12y = 12$

A) $(9, 1)$ B) $(-3, 1)$

C) $(3, 1)$ D) $(0, 1)$

59) Solve the below system of equations by using the method of elimination

$3x + 3y = -6$
$7x + 15y = 34$

A) $(6, 3)$ B) $(-8, -6)$

C) $(-8, 6)$ D) $(3, 6)$

60) Solve the below system of equations by using the method of elimination

$x + 12y = -3$
$-3x - 36y = 27$

A) $(-5, 12)$ B) No solution

C) $(5, -12)$ D) $(-10, 12)$

61) Solve the below system of equations by using the method of elimination

$6x + 3y = -3$
$-8x - 9y = 9$

A) $(-3, 1)$ B) $(0, 1)$

C) $(0, -1)$ D) $(-1, 0)$

62) Solve the below system of equations by using the method of elimination

$-6x - 5y = 11$
$18x + 11y = -5$

A) $(4, 7)$ B) $(4, -7)$

C) $(-8, -7)$ D) $(8, -7)$

63) Solve the below system of equations by using the method of elimination

$-5x - 12y = 5$
$8x - 24y = -8$

A) $(0, -3)$ B) No solution

C) $(0, -1)$ D) $(-1, 0)$

64) Solve the below system of equations by using the method of elimination

$7x + 24y = -3$
$-4x + 12y = -24$

A) $(1, 3)$ B) $(3, 1)$

C) $(3, -1)$ D) $(1, -3)$

Grade 8

Vol 1 Week 13 — System of equations

65) Solve the below system of equations by using the method of elimination

$$3x + 10y = -7$$
$$-9x - 30y = -3$$

A) $(-5, 5)$ B) $(-5, -5)$

C) $(5, -5)$ D) No solution

66) Solve the below system of equations by using the method of elimination

$$-5x + 16y = -28$$
$$x - 4y = 8$$

A) $(-3, 4)$ B) $(-4, -3)$

C) $(-4, 5)$ D) $(5, -4)$

67) Solve the below system of equations by using the method of elimination

$$-5x + 18y = 26$$
$$-9x + 6y = -6$$

A) $(-2, -2)$ B) $(-2, -11)$

C) $(-2, 2)$ D) $(2, 2)$

68) Solve the below system of equations by using the method of elimination

$$18x - 3y = 27$$
$$6x + 8y = -18$$

A) $(1, 3)$

B) $(1, -3)$

C) Infinite number of solutions

D) $(11, 3)$

69) Solve the below system of equations by using the method of elimination

$$-7x + 7y = -21$$
$$-2x + 14y = -30$$

A) No solution   B) $(1, 2)$

C) $(1, -2)$ D) $(2, 2)$

70) Solve the below system of equations by using the method of elimination

$$-12x - 6y = 12$$
$$24x + 9y = 0$$

A) $(3, -1)$ B) $(3, 1)$

C) $(3, -8)$ D) $(3, 7)$

71) Solve the below system of equations by using the method of elimination

$$-16x - 7y = -5$$
$$-8x + 3y = 17$$

A) $(11, -3)$ B) $(11, 3)$

C) $(-1, 3)$ D) $(-11, 3)$

72) Solve the below system of equations by using the method of elimination

$$18x + 3y = -21$$
$$-9x + 5y = 4$$

A) Infinite number of solutions

B) $(1, -1)$

C) $(-1, -1)$

D) $(-1, 1)$

Grade 8

Vol 1 Week 13 System of equations

73) Solve the below system of equations by using the method of elimination

$$8x + y = -10$$
$$16x - 5y = -6$$

A) $(-9, 2)$ B) $(-2, -1)$

C) $(-1, -2)$ D) $(9, 2)$

74) Solve the below system of equations by using the method of elimination

$$-11x - 5y = 4$$
$$-5x - y = 12$$

A) $(-9, -4)$ B) $(-4, 8)$

C) No solution   D) $(-4, 2)$

75) Solve the below system of equations by using the method of elimination

$$-3x - 2y = 31$$
$$-2x + 11y = 33$$

A) $(1, -8)$ B) $(-8, 1)$

C) $(-8, -6)$ D) $(-11, 1)$

76) Solve the below system of equations by using the method of elimination

$$40x + 8y = 32$$
$$30x + 6y = 24$$

A) No solution

B) Infinite number of solutions

C) $(2, -5)$

D) $(-5, 2)$

77) Solve the below system of equations by using the method of elimination

$$5x - 5y = -5$$
$$10x - 2y = 30$$

A) $(4, 5)$

B) $(-4, -5)$

C) Infinite number of solutions

D) $(4, -5)$

78) Solve the below system of equations by using the method of elimination

$$-5x - 7y = 35$$
$$-3x - 2y = 10$$

A) $(-5, 0)$

B) $(0, -5)$

C) Infinite number of solutions

D) $(5, 0)$

79) Solve the below system of equations by using the method of elimination

$$-7x - 3y = 33$$
$$-2x - 4y = -22$$

A) $(14, 10)$ B) $(-9, -10)$

C) $(10, 14)$ D) $(-9, 10)$

80) Solve the below system of equations by using the method of elimination

$$11x + 13y = 8$$
$$-4x - 2y = 8$$

A) No solution   B) $(-4, -5)$

C) $(-4, 4)$ D) $(-5, -4)$

Grade 8

Vol 1 Week 13 System of equations

81) Solve the below system of equations by using the method of elimination
$$6x - 6y = 12$$
$$-9x + 7y = 0$$
A) $(-9, 2)$
B) $(-7, -9)$
C) Infinite number of solutions
D) $(9, 2)$

82) Solve the below system of equations by using the method of elimination
$$-4x + 8y = -36$$
$$6x - 5y = -9$$
A) No solution
B) $(-9, -9)$
C) $(9, -9)$
D) $(-9, 9)$

83) Solve the below system of equations by using the method of elimination
$$10x + 13y = -35$$
$$4x + 2y = 2$$
A) $(14, -2)$
B) $(3, 5)$
C) $(3, -5)$
D) $(-2, 14)$

84) Solve the below system of equations by using the method of elimination
$$24x + 8y = -8$$
$$-30x - 10y = 10$$
A) Infinite number of solutions
B) $(-7, -11)$
C) $(-7, 13)$
D) $(13, -7)$

85) Solve the below system of equations by using the method of elimination
$$8x - 3y = 8$$
$$3x - 4y = 3$$
A) $(0, 11)$
B) $(-11, 0)$
C) $(-1, 0)$
D) $(1, 0)$

86) Solve the below system of equations by using the method of elimination
$$8x + 5y = -41$$
$$10x + 3y = -9$$
A) $(5, 13)$
B) $(-5, 13)$
C) $(3, -13)$
D) $(13, -5)$

87) Solve the below system of equations by using the method of elimination
$$-5x + 9y = -15$$
$$11x - 2y = 33$$
A) $(0, -3)$
B) $(3, 0)$
C) $(0, 3)$
D) $(-3, 0)$

88) Solve the below system of equations by using the method of elimination
$$5x + 4y = -33$$
$$11x - 13y = -29$$
A) $(-5, -2)$
B) $(-5, -9)$
C) $(-9, -5)$
D) $(-11, -9)$

Grade 8

Vol 1, Week 13 — System of equations

89) Solve the below system of equations by using the method of elimination

$$-5x - 6y = -37$$
$$-3x - 5y = -18$$

A) $(11, -3)$ B) $(-3, -11)$

C) $(-3, 11)$ D) $(-11, -3)$

90) Solve the below system of equations by using the method of elimination

$$10x + 20y = -24$$
$$-4x - 8y = 4$$

A) $(-12, 2)$ B) $(-2, -12)$

C) No solution   D) $(-2, 10)$

91) Solve the below system of equations by using the method of elimination

$$10x - 7y = -7$$
$$6x + 13y = 13$$

A) No solution   B) $(1, -12)$

C) $(1, 12)$ D) $(0, 1)$

92) Solve the below system of equations by using the method of elimination

$$-13x + 7y = 4$$
$$-11x + 10y = 36$$

A) $(-4, 6)$ B) $(-4, 8)$

C) $(4, 8)$ D) $(4, -8)$

93) When you reverse the digits in a certain two-digit number you decrease its value by 54. What is the number if the sum of its digits is 10?

94) When you reverse the digits in a certain two-digit number you decrease its value by 9. Find the number if the sum of its digits is 15.

95) The sum of the digits of a certain two-digit number is 11. When you reverse its digits you increase the number by 27. What is the number?

96) When you reverse the digits in a certain two-digit number you decrease its value by 36. Find the number if the sum of its digits is 6.

97) When you reverse the digits in a certain two-digit number you decrease its value by 63. What is the number if the sum of its digits is 7?

Grade 8

Vol 1 Week 13 System of equations

98) When you reverse the digits in a certain two-digit number you increase its value by 72. What is the number if the sum of its digits is 10?

99) The sum of the digits of a certain two-digit number is 5. When you reverse its digits you decrease the number by 27. Find the number.

100) When you reverse the digits in a certain two-digit number you increase its value by 54. Find the number if the sum of its digits is 10.

101) When you reverse the digits in a certain two-digit number you decrease its value by 45. What is the number if the sum of its digits is 5?

102) When you reverse the digits in a certain two-digit number you increase its value by 27. Find the number if the sum of its digits is 9.

103) The sum of the digits of a certain two-digit number is 13. When you reverse its digits you decrease the number by 9. Find the number.

104) When you reverse the digits in a certain two-digit number you decrease its value by 9. Find the number if the sum of its digits is 11.

105) A farmhouse shelters 14 animals. Some are horses and some are geese. Altogether there are 48 legs. How many of each animal are there?

A) 10 geese and 4 horses

B) 12 geese and 2 horses

C) 4 geese and 10 horses

D) 11 geese and 3 horses

106) There are 15 animals in the barn. Some are chickens and some are sheep. There are 50 legs in all. How many of each animal are there?

A) 5 chickens and 10 sheep

B) 12 chickens and 3 sheep

C) 13 chickens and 2 sheep

D) 11 chickens and 4 sheep

Grade 8

Vol 1
Week 13
System of equations

107) A farmhouse shelters 23 animals. Some are pigs and some are chickens. Altogether there are 84 legs. How many of each animal are there ?

A) 4 chickens and 19 pigs

B) 20 chickens and 2 pigs

C) 21 chickens and 2 pigs

D) 20 chickens and 3 pigs

108) There are 12 animals in the barn. Some are ducks and some are horses. There are 34 legs in all. How many of each animal are there ?

A) 8 ducks and 4 horses

B) 9 ducks and 3 horses

C) 7 ducks and 5 horses

D) 10 ducks and 2 horses

109) A farmhouse shelters 26 animals. Some are sheep and some are ducks. Altogether there are 88 legs. How many of each animal are there ?

A) 22 ducks and 4 sheep

B) 23 ducks and 3 sheep

C) 8 ducks and 18 sheep

D) 24 ducks and 2 sheep

110) There are 12 animals in the barn. Some are geese and some are pigs. There are 42 legs in all. How many of each animal are there ?

A) 11 geese and 3 pigs

B) 10 geese and 2 pigs

C) 9 geese and 3 pigs

D) 3 geese and 9 pigs

111) A farmhouse shelters 12 animals. Some are cows and some are chickens. Altogether there are 40 legs. How many of each animal are there ?

A) 8 chickens and 4 cows

B) 10 chickens and 2 cows

C) 9 chickens and 3 cows

D) 4 chickens and 8 cows

112) There are 16 animals in the barn. Some are ducks and some are goats. There are 58 legs in all. How many of each animal are there ?

A) 13 ducks and 3 goats

B) 16 ducks and 2 goats

C) 14 ducks and 2 goats

D) 3 ducks and 13 goats

Grade 8

Vol 1 Week 13 System of equations

113) There are 10 animals in the barn. Some are ducks and some are sheep. There are 30 legs in all. How many of each animal are there?

   A) 7 ducks and 3 sheep

   B) 9 ducks and 2 sheep

   C) 5 ducks and 5 sheep

   D) 8 ducks and 2 sheep

114) A farmhouse shelters 15 animals. Some are cows and some are chickens. Altogether there are 42 legs. How many of each animal are there?

   A) 13 chickens and 2 cows

   B) 12 chickens and 2 cows

   C) 9 chickens and 6 cows

   D) 14 chickens and 3 cows

115) There are 15 animals. Some are ducks and some are pigs. There are 52 legs in all. How many of each animal are there?

   A) 12 ducks and 3 pigs

   B) 14 ducks and 3 pigs

   C) 13 ducks and 2 pigs

   D) 4 ducks and 11 pigs

116) A farmhouse shelters 18 animals. Some are buffalo and some are ducks. Altogether there are 62 legs. How many of each animal are there?

   A) 5 ducks and 13 buffalos

   B) 15 ducks and 3 buffalos

   C) 17 ducks and 3 buffalos

   D) 16 ducks and 2 buffalos

117) There are 24 animals in the barn. Some are geese and some are buffalo. There are 84 legs in all. How many of each animal are there?

   A) 6 geese and 18 buffalo

   B) 19 geese and 3 buffalo

   C) 8 geese and 16 buffalo

   D) 22 geese and 2 buffalo

118) There are 27 animals in the barn. Some are geese and some are sheep. There are 94 legs in all. How many of each animal are there?

   A) 22 geese and 5 sheep

   B) 24 geese and 3 sheep

   C) 25 geese and 2 sheep

   D) 7 geese and 20 sheep

Grade 8

Vol 1 Week 13 System of equations

119) A farmhouse shelters 13 animals. Some are pigs and some are chickens. Altogether there are 48 legs in all. How many of each animal are there?

   A) 9 chickens and 4 pigs

   B) 8 chickens and 5 pigs

   C) 2 chickens and 11 pigs

   D) 11 chickens and 2 pigs

Tracy bought 11 games for a total of $293. Game A cost $25 and Game B cost $28.

120) How many number of game A's did she buy?

   A) 5   B) 8

   C) 7   D) 9

121) How many number of game B's did she buy?

   A) 2   B) 3

   C) 4   D) 6

Jack spent $610 on story books. Fiction books costs $50 and non-fiction books costs $60. He bought a total of 11 story books.

122) How many number of fiction books did he buy?

   A) 5 Fiction books   B) 9 Fiction books

   C) 8 Fiction books   D) 8 Fiction books

123) How many number of non fiction books did he buy?

   A) 6 non-fiction books

   B) 2 non-fiction books

   C) 3 non-fiction books

   D) 2 non-fiction books

Lucy bought 10 dresses for a total of $204. Formal dress cost $26 and casual dress cost $12.

124) How many number of formal dresses did she buy?

   A) 6 formal dresses

   B) 7 formal dresses

   C) 4 formal dresses

   D) 7 formal dresses

125) How many number of casual dresses did she buy?

   A) 6 casual dresses

   B) 3 casual dresses

   C) 2 casual dresses

   D) 2 casual dresses

Grade 8

Vol 1 Week 13 System of equations

Julia spent $40 on snacks. Snack A cost $5 and snack B cost $10. She bought a total of 6 snacks.

126) How many number of snack A did she buy?

A) 5 snack A   B) 4 snack A

C) 2 snack A   D) 3 snack A

127) How many number of snack B did she buy?

A) 3 snack B   B) 3 snack B

C) 4 snack B   D) 2 snack B

Grade 8

Vol 1 Week 14 System of equations

1) Solve the below system of equations by using the method of substitution

$x - y = -5$
$-2x - 7y = 19$

A) $(-6, -1)$

B) Infinite number of solutions

C) $(-6, -10)$

D) $(-6, 10)$

2) Solve the below system of equations by using the method of substitution

$x + 6y = -3$
$12x - 2y = -36$

A) $(-3, 0)$ B) $(0, 4)$
C) $(0, -3)$ D) $(3, 0)$

3) Solve the below system of equations by using the method of substitution

$-12x - 3y = -6$
$-7x + y = 24$

A) $(-2, 10)$ B) $(11, -10)$
C) $(7, -10)$ D) $(-2, -10)$

4) Solve the below system of equations by using the method of substitution

$9x + 10y = 12$
$-x + y = -14$

A) $(-8, -2)$ B) $(8, -6)$
C) $(-8, 2)$ D) $(8, -2)$

5) Solve the below system of equations by using the method of substitution

$-2x - 6y = -10$
$x + 11y = -3$

A) $(-3, 6)$ B) $(8, -1)$
C) $(-3, -1)$ D) No solution

6) Solve the below system of equations by using the method of substitution

$-11x + y = -1$
$5x - 4y = 4$

A) $(1, 2)$ B) $(1, -10)$
C) $(0, 1)$ D) $(0, -1)$

7) Solve the below system of equations by using the method of substitution

$-2x - 5y = 27$
$x - 5y = 9$

A) No solution   B) $(-6, -3)$
C) $(10, -3)$ D) $(4, 6)$

8) Solve the below system of equations by using the method of substitution

$-3x + 2y = 28$
$-3x + y = 32$

A) $(9, -4)$ B) $(-9, -4)$
C) $(-12, -4)$ D) $(-4, -9)$

Grade 8 — Vol 1, Week 14 — System of equations

9) Solve the below system of equations by using the method of substitution

$$9x + 3y = 1$$
$$3x + y = 7$$

A) No solution B) $(1, 10)$
C) $(-1, -10)$ D) $(-1, 10)$

10) Solve the below system of equations by using the method of substitution

$$3x - 5y = 0$$
$$x - 2y = 0$$

A) $(0, 10)$ B) $(-10, 0)$
C) $(0, 0)$ D) $(0, -10)$

11) Solve the below system of equations by using the method of substitution

$$x - 3y = 4$$
$$-2x + 8y = -10$$

A) $(-1, 5)$ B) $(1, 5)$
C) $(1, -1)$ D) $(-1, -5)$

12) Solve the below system of equations by using the method of substitution

$$-6x - 3y = 3$$
$$2x + y = -1$$

A) $(-12, -4)$

B) $(-12, 4)$

C) $(4, -12)$

D) Infinite number of solutions

13) Solve the below system of equations by using the method of substitution

$$y = -11$$
$$4x - y = -29$$

A) $(-10, 5)$ B) $(-10, -3)$
C) $(-10, 11)$ D) $(-10, -11)$

14) Solve the below system of equations by using the method of substitution

$$-6x - 2y = -24$$
$$y = 0$$

A) $(4, -5)$ B) $(-4, -5)$
C) $(4, 0)$ D) $(4, 5)$

15) Solve the below system of equations by using the method of substitution

$$12x + 6y = -30$$
$$-6x + 8y = -18$$

A) $(-1, -3)$ B) $(-1, 3)$
C) $(1, 3)$ D) $(1, -3)$

16) Solve the below system of equations by using the method of substitution

$$9x + y = -14$$
$$11x + 5y = -36$$

A) $(-9, -1)$ B) $(-9, 3)$
C) $(-1, -5)$ D) No solution

Grade 8

**Vol 1
Week 14
System of equations**

17) Solve the below system of equations by using the method of substitution

$4x - 11y = 20$
$x + 9y = 5$

A) $(-5, 0)$

B) Infinite number of solutions

C) No solution

D) $(5, 0)$

18) Solve the below system of equations by using the method of substitution

$-x + y = 0$
$5x - 2y = -27$

A) $(9, -9)$ B) $(-9, -9)$

C) $(-9, 1)$ D) $(-9, -8)$

19) Solve the below system of equations by using the method of substitution

$9x + 11y = 14$
$x - y = -14$

A) No solution

B) $(-7, 7)$

C) Infinite number of solutions

D) $(-7, -7)$

20) Solve the below system of equations by using the method of substitution

$-3x + 21y = -99$
$x - 7y = 33$

A) $(-12, 2)$

B) $(-12, -2)$

C) $(12, -2)$

D) Infinite number of solutions

21) Solve the below system of equations by using the method of substitution

$-9x + 7y = -11$
$-11x + y = -21$

A) $(1, 2)$

B) Infinite number of solutions

C) $(1, -2)$

D) $(2, 1)$

22) Solve the below system of equations by using the method of substitution

$y = 10$
$3x + 2y = 8$

A) $(-10, 4)$ B) $(-4, -10)$

C) $(4, -10)$ D) $(-4, 10)$

23) Solve the below system of equations by using the method of substitution

$x + 4y = -5$
$8x + 7y = 35$

A) $(-2, 7)$ B) $(8, 7)$

C) $(-3, 7)$ D) $(7, -3)$

24) Solve the below system of equations by using the method of substitution

$y = -1$
$6x - 11y = -7$

A) $(3, 1)$ B) $(-1, 3)$

C) $(-3, -1)$ D) $(1, 3)$

Grade 8

Vol 1 Week 14 System of equations

25) Solve the below system of equations by using the method of substitution

$$x - 9y = -13$$
$$6x + 10y = -14$$

A) $(1, 4)$ B) No solution
C) $(4, 1)$ D) $(-4, 1)$

26) Solve the below system of equations by using the method of substitution

$$-12x - 12y = -36$$
$$-12x - 5y = 34$$

A) $(-8, 7)$ B) $(10, -7)$
C) $(-5, 7)$ D) $(-7, 10)$

27) Solve the below system of equations by using the method of substitution

$$10x - 3y = 2$$
$$-11x + 6y = 14$$

A) $(2, -6)$ B) $(-7, -2)$
C) $(-6, -2)$ D) $(2, 6)$

28) Solve the below system of equations by using the method of substitution

$$-2x + 10y = 8$$
$$4x - 20y = 4$$

A) No solution B) $(-3, -12)$
C) $(-3, 12)$ D) $(3, 12)$

29) Solve the below system of equations by using the method of substitution

$$x + 4y = -19$$
$$3x + 12y = -57$$

A) $(11, -4)$
B) $(-12, 4)$
C) $(-11, -4)$
D) Infinite number of solutions

30) Solve the below system of equations by using the method of substitution

$$7x + 11y = 11$$
$$-3x + 2y = 2$$

A) $(6, 8)$ B) $(0, 1)$
C) $(1, 0)$ D) $(-1, 0)$

31) Solve the below system of equations by using the method of substitution

$$-4x - 8y = 36$$
$$y = -7$$

A) $(5, -7)$
B) $(-7, -3)$
C) $(-7, 5)$
D) Infinite number of solutions

32) Solve the below system of equations by using the method of substitution

$$4x + 5y = 24$$
$$8x + 4y = 0$$

A) No solution B) $(-8, 4)$
C) $(-4, -8)$ D) $(-4, 8)$

Grade 8

Vol 1
Week 14
System of equations

33) Solve the below system of equations by using the method of substitution

$-10x - 5y = -10$
$-2x + 2y = 28$

A) $(-10, -4)$ B) $(-4, 10)$
C) $(4, 10)$ D) $(10, 4)$

34) Solve the below system of equations by using the method of substitution

$-9x + 7y = 24$
$5x - y = -22$

A) $(-5, -3)$ B) $(3, -5)$
C) $(-5, 3)$ D) $(5, -3)$

35) Solve the below system of equations by using the method of substitution

$-12x - 12y = 12$
$-5x - 9y = -3$

A) $(-3, 2)$ B) $(-3, -12)$
C) $(-3, 12)$ D) $(3, -12)$

36) Solve the below system of equations by using the method of substitution

$9x + 5y = 0$
$-2x - 2y = -8$

A) $(-8, 9)$

B) Infinite number of solutions

C) $(8, 5)$

D) $(-5, 9)$

37) Solve the below system of equations by using the method of substitution

$-5x + 3y = 0$
$12x - y = -31$

A) $(-7, 3)$ B) $(-5, -3)$
C) $(-7, -3)$ D) $(-3, -5)$

38) Solve the below system of equations by using the method of substitution

$-6x - 5y = -32$
$12x + 3y = 36$

A) $(2, 4)$ B) $(-12, 1)$
C) $(-12, -8)$ D) $(-4, 2)$

39) Solve the below system of equations by using the method of substitution

$4x + 11y = 29$
$-8x - 22y = -58$

A) No solution

B) $(-12, -5)$

C) Infinite number of solutions

D) $(-12, -10)$

40) Solve the below system of equations by using the method of substitution

$-9x + 4y = 23$
$y = -1$

A) $(-3, -2)$ B) $(-2, -6)$
C) $(-3, -1)$ D) $(-6, -2)$

Grade 8

Vol 1
Week 14
System of equations

41) Solve the below system of equations by using the method of substitution

$$3x + 12y = 96$$
$$-x - 4y = -32$$

A) $(-5, 10)$

B) Infinite number of solutions

C) No solution

D) $(-5, -9)$

42) Solve the below system of equations by using the method of substitution

$$-4x - 9y = -33$$
$$2x + 4y = 14$$

A) $(3, 12)$ B) $(3, 5)$

C) $(-3, 12)$ D) $(-3, 5)$

43) The sum of two numbers is 24. Their difference is 2. What are the numbers?

A) 18 and 19   B) 11 and 13

C) 17 and 8   D) 14 and 7

44) The difference of two numbers is 3. Their sum is 25. What are the numbers?

A) 11 and 14   B) 12 and 19

C) 14 and 13   D) 18 and 19

45) Find the value of two numbers if their sum is 20 and their difference is 4.

A) 7 and 14   B) 12 and 5

C) 5 and 18   D) 8 and 12

46) The sum of two numbers is 26. Their difference is 2. Find the numbers.

A) 14 and 10   B) 16 and 16

C) 16 and 22   D) 12 and 14

47) The sum of two numbers is 22. Their difference is 6. What are the numbers?

A) 4 and 12   B) 8 and 14

C) 13 and 21   D) 5 and 13

48) Find the value of two numbers if their sum is 18 and their difference is 2.

A) 4 and 12   B) 6 and 14

C) 7 and 14   D) 8 and 10

48) Find the value of two numbers if their sum is 18 and their difference is 2.

A) 4 and 12   B) 6 and 14

C) 7 and 14   D) 8 and 10

Grade 8

Vol 1 Week 14 System of equations

49) Find the value of two numbers if their sum is 15 and their difference is 1.

   A) 3 and 13   B) 10 and 7

   C) 8 and 6   D) 7 and 8

50) Find the value of two numbers if their sum is 24 and their difference is 4.

   A) 8 and 12   B) 10 and 14

   C) 4 and 17   D) 4 and 17

51) Find the value of two numbers if their sum is 20 and their difference is 2.

   A) 11 and 7   B) 11 and 7

   C) 11 and 4   D) 5 and 14

52) Find the value of two numbers if their sum is 23 and their difference is 3.

   A) 4 and 6   B) 7 and 20

   C) 13 and 21   D) 10 and 13

53) The difference of two numbers is 5. Their sum is 23. What are the numbers?

   A) 14 and 7   B) 12 and 21

   C) 9 and 14   D) 6 and 17

54) Find the value of two numbers if their sum is 23 and their difference is 1.

   A) 12 and 10   B) 18 and 11

   C) 8 and 10   D) 11 and 12

55) The difference of two numbers is 1. Their sum is 19. Find the numbers.

   A) 8 and 14   B) 8 and 14

   C) 9 and 10   D) 7 and 5

56) Find the value of two numbers if their sum is 27 and their difference is 1.

   A) 15 and 22   B) 10 and 15

   C) 8 and 19   D) 13 and 14

57) The difference of two numbers is 2. Their sum is 22. What are the numbers?

   A) 15 and 6   B) 12 and 13

   C) 14 and 15   D) 10 and 12

58) The difference of two numbers is 5. Their sum is 19. Find the numbers.

   A) 9 and 16   B) 9 and 10

   C) 7 and 12   D) 3 and 8

Grade 8

Vol 1 Week 14 System of equations

59) The sum of two numbers is 16. Their difference is 2. Find the numbers.

   A) 7 and 9   B) 8 and 13

   C) 8 and 14   D) 3 and 11

60) The sum of two numbers is 21. Their difference is 3. What are the numbers?

   A) 14 and 8   B) 11 and 17

   C) 9 and 12   D) 5 and 8

61) Find the value of two numbers if their sum is 19 and their difference is 3.

   A) 3 and 18   B) 12 and 16

   C) 3 and 8   D) 8 and 11

62) The sum of two numbers is 25. Their difference is 1. Find the numbers.

   A) 12 and 13   B) 17 and 5

   C) 18 and 21   D) 5 and 19

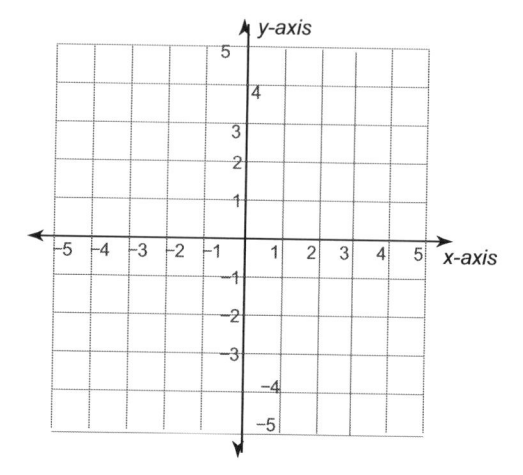

63) Find the solution of the system of equations given below by graphing

$$3x - y = 4$$
$$4x + y = 3$$

A) No solution   B) $(1, -1)$

C) $(1, 1)$ D) $(-1, 1)$

X	Y
0	
	0

Grade 8

Vol 1 Week 14 System of equations

64) Find the solution of the system of equations given below by graphing

$$y = \frac{7}{3}x - 4$$

$$y = \frac{7}{3}x - 3$$

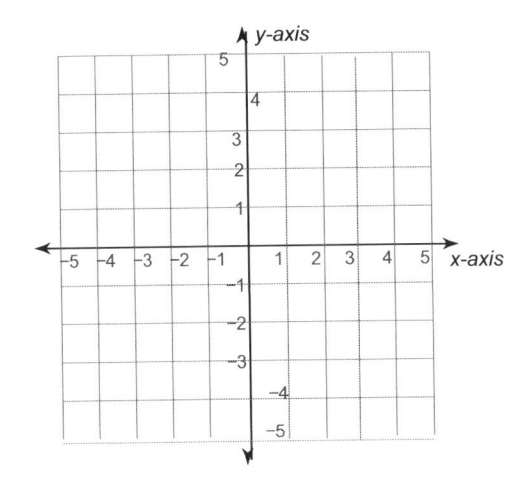

A) $(5, 3)$ B) $(1, 3)$

C) $(5, -3)$ D) No solution

X	Y
0	
	0

65) Find the solution of the system of equations given below by graphing

$$y = -\frac{1}{2}x + 2$$

$$y = -3x - 3$$

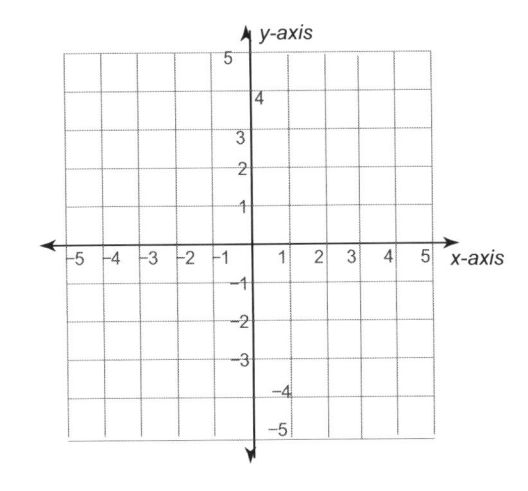

A) $(-2, 3)$

B) $(-2, 1)$

C) $(2, 1)$

D) Infinite number of solutions

X	Y
0	
	0

66) Find the solution of the system of equations given below by graphing

$$y = -\frac{1}{4}x + 4$$

$$y = \frac{3}{2}x - 3$$

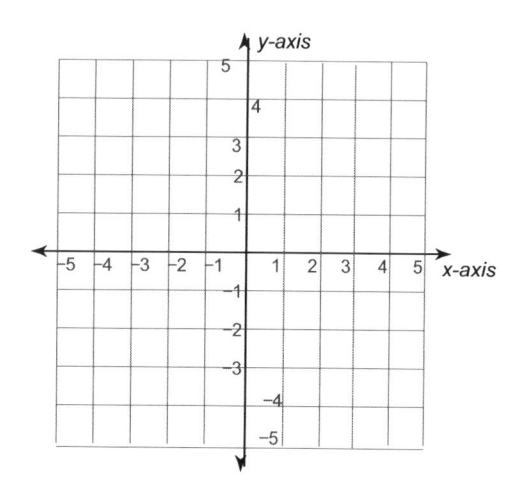

A) $(4, -3)$ B) $(4, 3)$

C) $(-3, 4)$ D) $(-2, 3)$

X	Y
0	
	0

67) Find the solution of the system of equations given below by graphing

$$y = 3$$

$$y = -2x + 1$$

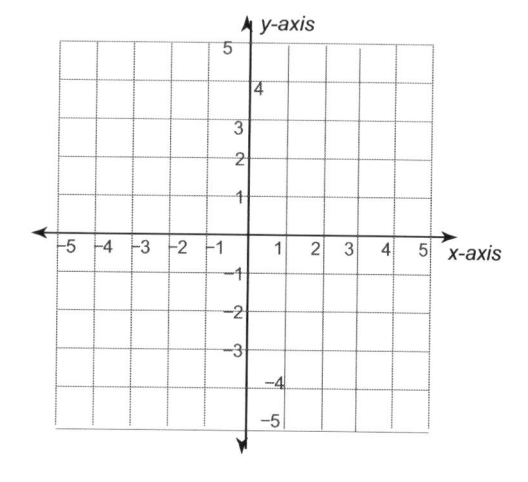

A) $(-1, 3)$ B) $(-3, -1)$

C) $(2, -1)$ D) $(3, 1)$

X	Y
0	
	0

Grade 8

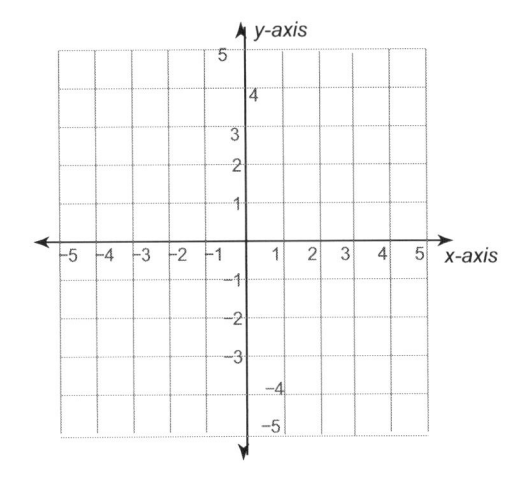

Vol 1 Week 14 System of equations

68) Find the solution of the system of equations given below by graphing

$$y = 2x + 1$$
$$y = -x - 2$$

A) $(4, -1)$ B) $(-1, -1)$

C) $(-1, -4)$ D) $(-4, -1)$

X	Y
0	
	0

69) Find the solution of the system of equations given below by graphing

$$y = \frac{2}{3}x - 2$$
$$y = 2x + 2$$

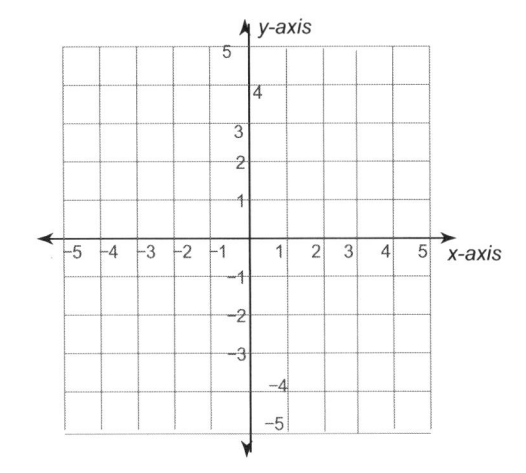

A) $(-3, -4)$ B) $(3, -4)$

C) $(4, -3)$ D) $(-4, 3)$

X	Y
0	
	0

Grade 8

**Vol 1
Week 14
System of equations**

70) Find the solution of the system of equations given below by graphing

$$y = -\frac{1}{2}x - 4$$

$$y = \frac{5}{4}x + 3$$

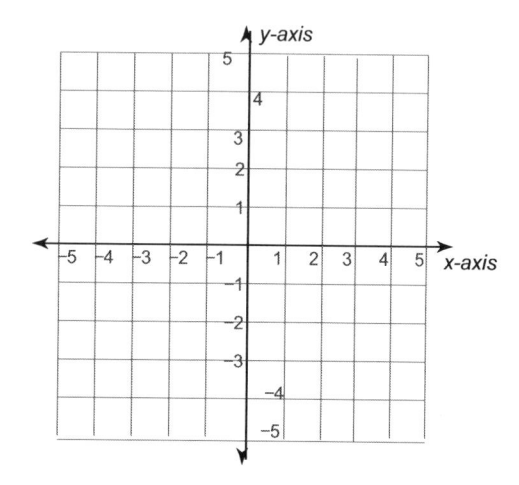

A) $(5, -2)$ B) $(-4, -2)$

C) $(5, 2)$ D) $(2, -5)$

X	Y
0	
	0

71) Find the solution of the system of equations given below by graphing

$$y = -\frac{1}{2}x - 3$$

$$y = -\frac{5}{2}x + 1$$

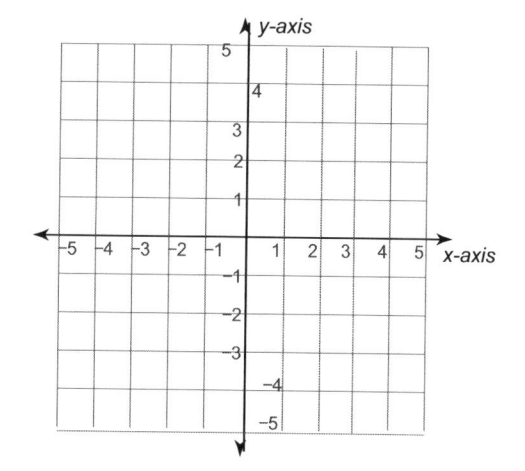

A) $(-4, 2)$ B) $(2, -4)$

C) $(-4, 3)$ D) $(-4, -3)$

X	Y
0	
	0

230

Grade 8

Vol 1 Week 14 System of equations

72) Find the solution of the system of equations given below by graphing

$$y = -\frac{7}{3}x - 4$$

$$y = -\frac{2}{3}x + 1$$

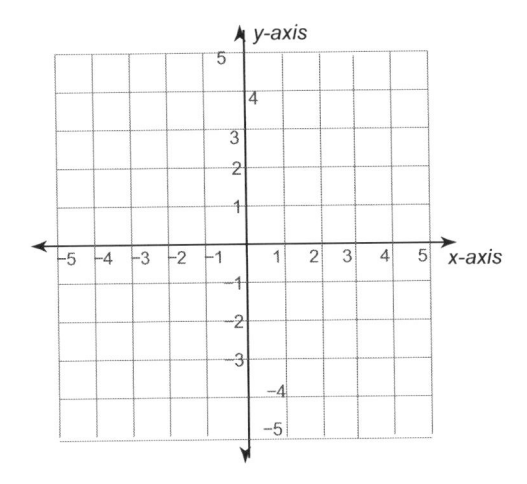

A) $(-3, 3)$

B) $(3, 3)$

C) No solution

D) Infinite number of solutions

X	Y
0	
	0

73) Find the solution of the system of equations given below by graphing

$$y = -\frac{1}{2}x - 4$$

$$y = \frac{7}{2}x + 4$$

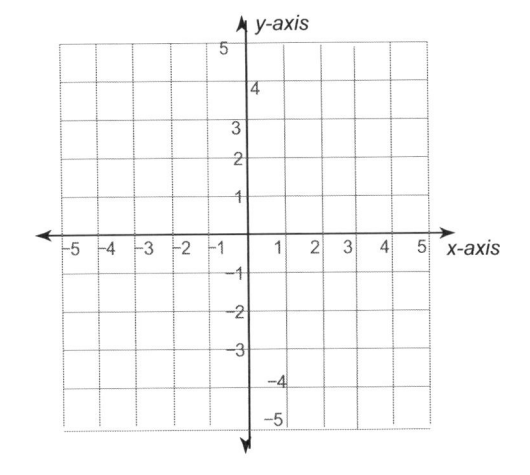

A) $(-2, -3)$ B) $(-2, 2)$

C) $(2, 2)$ D) $(5, 2)$

X	Y
0	
	0

74) Find the solution of the system of equations given below by graphing

$$y = \frac{7}{4}x - 4$$

$$y = \frac{1}{2}x + 1$$

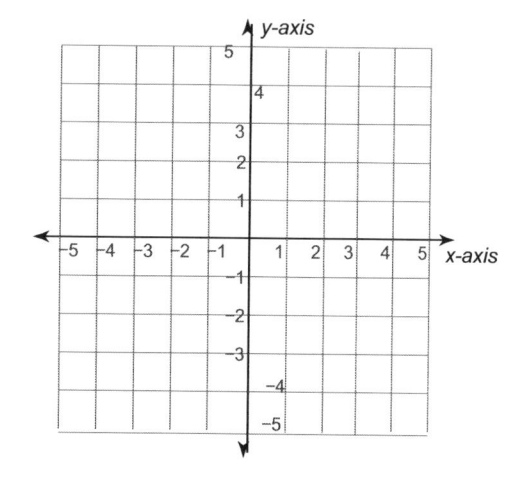

A) $(-3, 4)$ B) No solution

C) $(4, 3)$ D) $(4, -3)$

X	Y
0	
	0

75) Find the solution of the system of equations given below by graphing

$$y = -2x + 2$$

$$y = -2x - 4$$

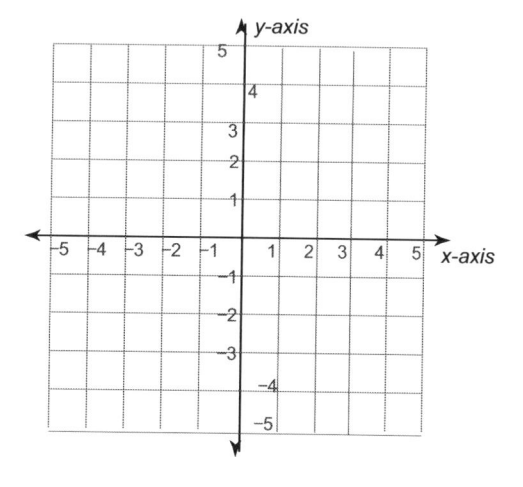

A) $(1, 2)$

B) No solution

C) $(-1, -2)$

D) Infinite number of solutions

X	Y
0	
	0

Grade 8

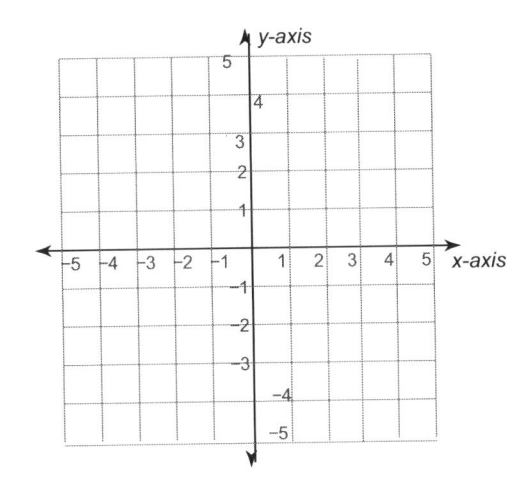

Vol 1
Week 14
System of equations

76) Find the solution of the system of equations given below by graphing

$$y = 2x + 4$$
$$y = 2x - 2$$

A) No solution B) (5, 1)

C) (−4, 1) D) (−5, 1)

X	Y
0	
	0

77) Find the solution of the system of equations given below by graphing

$$y = -\frac{3}{2}x + 3$$
$$y = -\frac{1}{4}x - 2$$

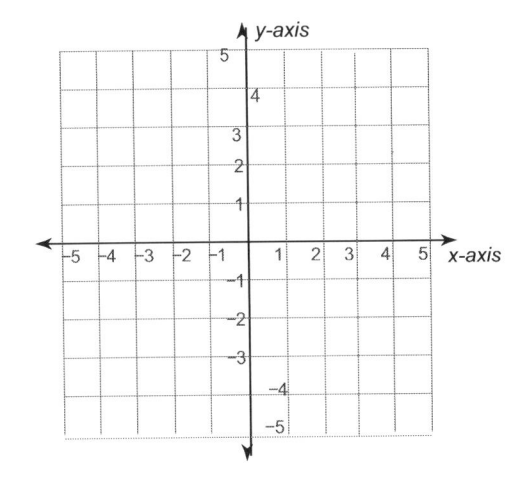

A) (−3, −3) B) (−3, 4)

C) (4, −3) D) (5, −3)

X	Y
0	
	0

Grade 8

Vol 1
Week 14
System of equations

78) Find the solution of the system of equations given below by graphing

$$y = -\frac{2}{3}x + 3$$

$$y = -\frac{2}{3}x - 1$$

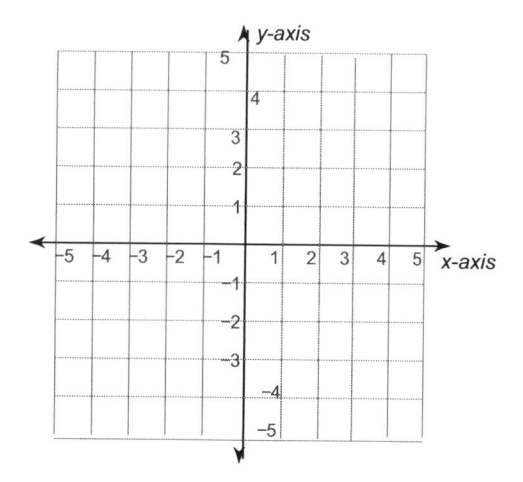

A) $(2, -5)$　　B) $(-5, 2)$

C) No solution　　D) $(-5, -2)$

X	Y
0	
	0

79) Find the solution of the system of equations given below by graphing

$$y = \frac{7}{2}x - 3$$

$$y = \frac{7}{2}x + 3$$

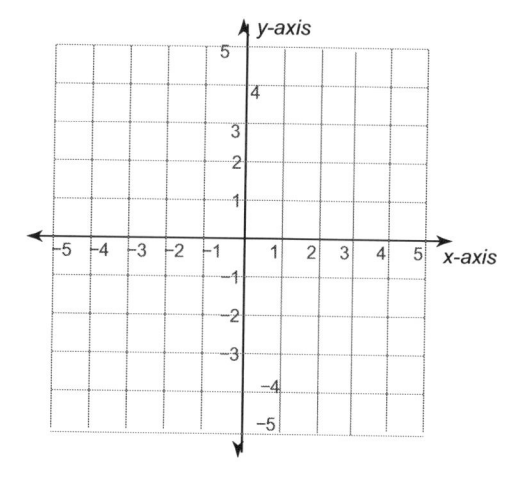

A) Infinite number of solutions

B) $(-1, 5)$

C) $(-1, -5)$

D) No solution

X	Y
0	
	0

Grade 8

Vol 1 Week 14 System of equations

80) Find the solution of the system of equations given below by graphing

$$y = -\frac{1}{2}x + 2$$

$$y = -\frac{3}{2}x - 2$$

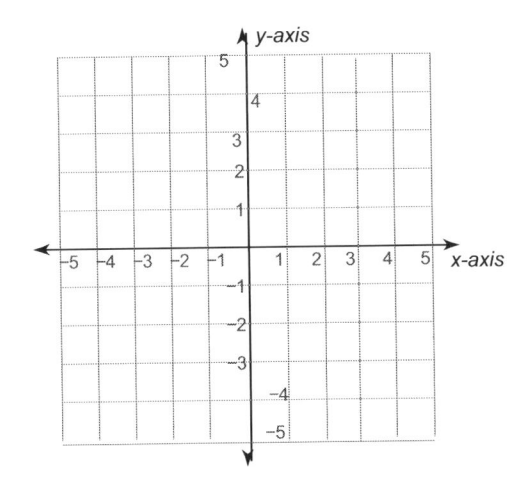

A) (4, 4)   B) No solution

C) (3, 4)   D) (−4, 4)

X	Y
0	
	0

81) Find the solution of the system of equations given below by graphing

$$y = x - 4$$

$$y = -3x + 4$$

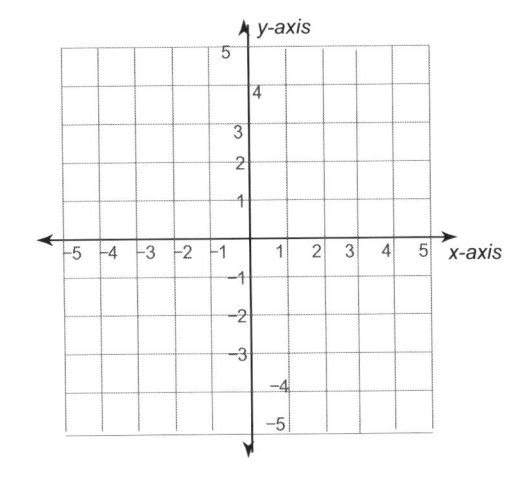

A) (5, −2)   B) (5, 2)

C) (2, −2)   D) (2, 5)

X	Y
0	
	0

Grade 8

Vol 1 Week 14 System of equations

82) Find the solution of the system of equations given below by graphing

$$y = \frac{3}{2}x + 2$$

$$y = -\frac{1}{2}x - 2$$

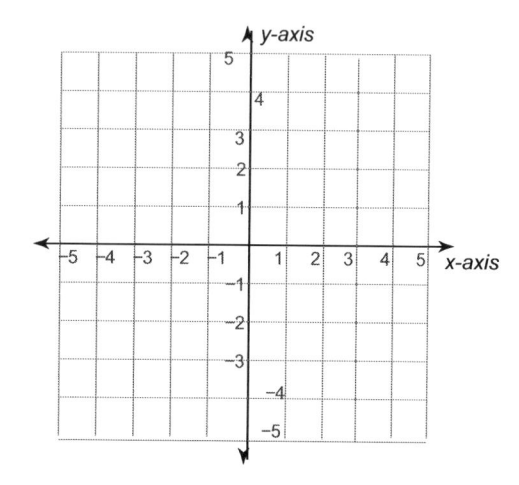

A) No solution   B) (2, 1)

C) (−2, −1)   D) (2, −1)

X	Y
0	
	0

83) Find the solution of the system of equations given below by graphing

$$y = \frac{1}{2}x + 3$$

$$y = \frac{5}{2}x - 1$$

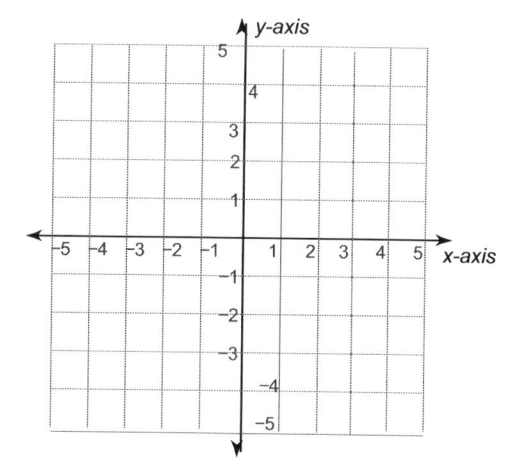

A) (2, −4)   B) (2, 4)

C) (−4, −2)   D) (−2, −4)

X	Y
0	
	0

Grade 8

Vol 1 Week 14 System of equations

84) Find the solution of the system of equations given below by graphing

$$y = \frac{1}{2}x - 4$$

$$y = -\frac{1}{2}x - 2$$

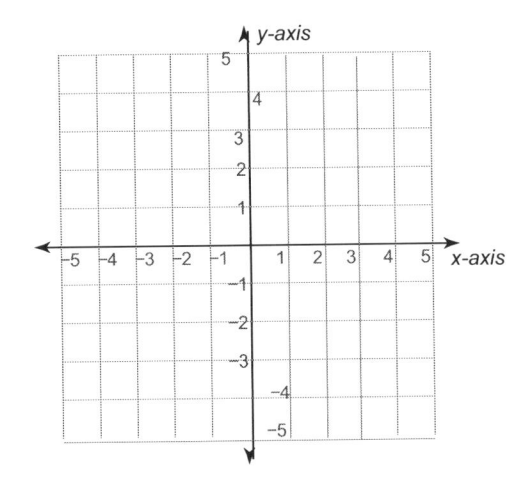

A) $(-2, -3)$ B) $(-2, 3)$

C) $(2, -3)$ D) $(-3, 2)$

X	Y
0	
	0

85) Find the solution of the system of equations given below by graphing

$$y = 2x + 4$$

$$y = \frac{1}{2}x - 2$$

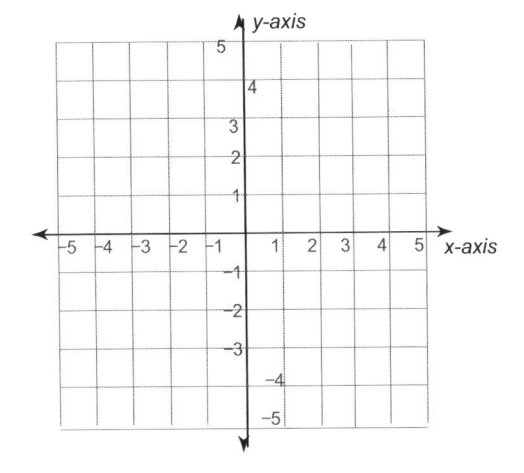

A) $(4, 4)$ B) $(-4, -4)$

C) $(4, -4)$ D) $(-4, 4)$

X	Y
0	
	0

Grade 8

Vol 1
Week 14
System of equations

86) Find the solution of the system of equations given below by graphing

$$y = \frac{1}{2}x + 2$$

$$y = \frac{7}{2}x - 4$$

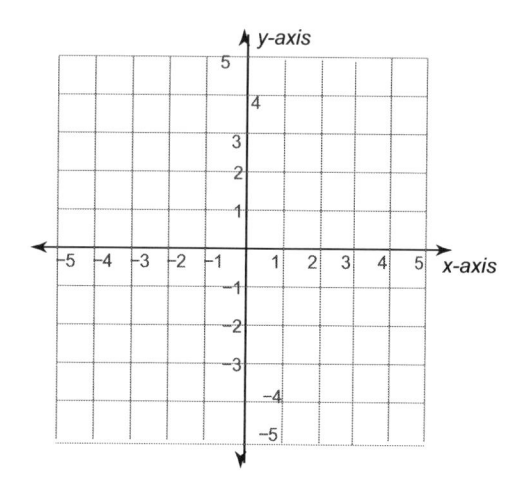

A) $(3, 2)$ B) $(-3, 2)$
C) $(4, -4)$ D) $(2, 3)$

X	Y
0	
	0

87) Find the solution of the system of equations given below by graphing

$$y = -x + 3$$

$$y = 3x - 1$$

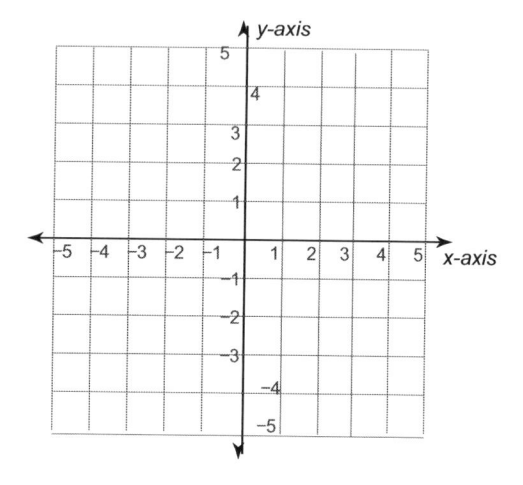

A) $(1, 2)$ B) $(-1, -3)$
C) $(-3, -1)$ D) $(-1, 2)$

X	Y
0	
	0

Grade 8

Vol 1　Week 14　System of equations

88) Find the solution of the system of equations given below by graphing

$$y = \frac{2}{3}x + 1$$

$$y = -\frac{1}{3}x + 4$$

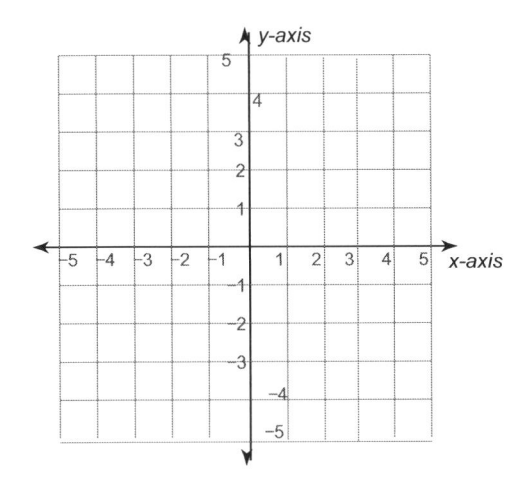

A) $(-3, 5)$　　B) $(3, 3)$

C) $(-3, -5)$　　D) $(3, 5)$

X	Y
0	
	0

89) Find the solution of the system of equations given below by graphing

$$y = \frac{1}{2}x + 3$$

$$y = -\frac{3}{4}x - 2$$

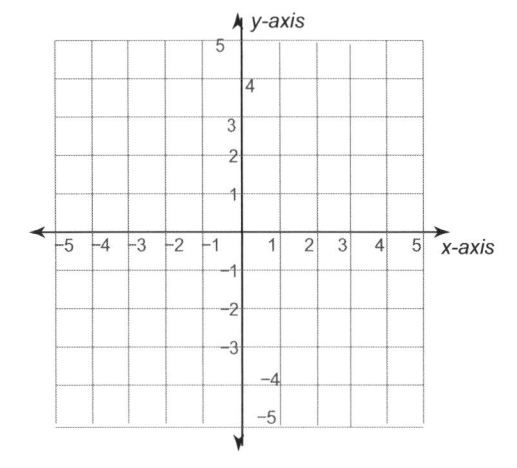

A) $(4, -4)$　　B) $(4, 1)$

C) $(-4, 1)$　　D) $(4, 3)$

X	Y
0	
	0

Grade 8

Vol 1
Week 14
System of equations

90) Find the solution of the system of equations given below by graphing

$$y = x - 3$$

$$y = -\frac{3}{4}x + 4$$

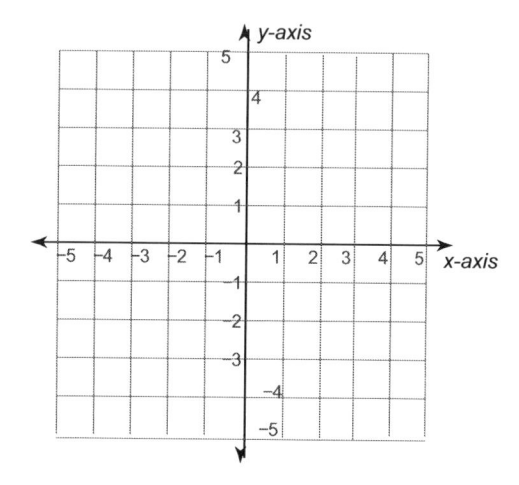

A) (4, 1) B) (–3, 1)
C) (–2, 1) D) (1, 1)

X	Y
0	
	0

91) Find the solution of the system of equations given below by graphing

$$y = -2x + 4$$

$$y = 4x - 2$$

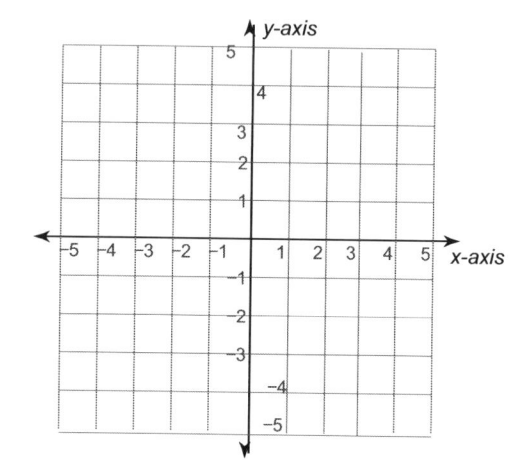

A) No solution B) (3, –1)
C) (1, 2) D) (1, 1)

X	Y
0	
	0

Grade 8

Vol 1
Week 14
System of equations

92) Find the solution of the system of equations given below by graphing

$$y = -\frac{1}{2}x + 4$$

$$y = \frac{5}{2}x - 2$$

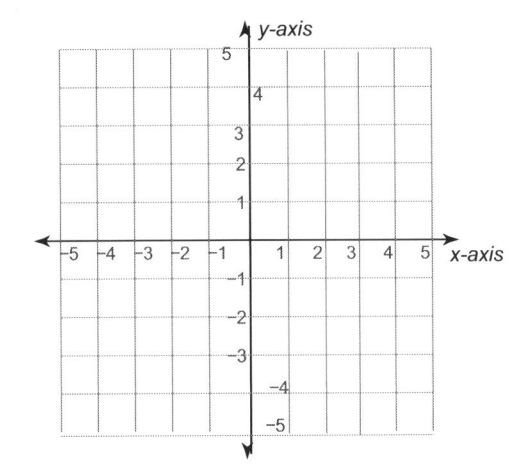

A) $(2, -1)$ B) $(2, 3)$
C) $(2, 1)$ D) $(-2, 1)$

X	Y
0	
	0

93) Find the solution of the system of equations given below by graphing

$$y = -\frac{7}{3}x - 3$$

$$y = -\frac{2}{3}x + 2$$

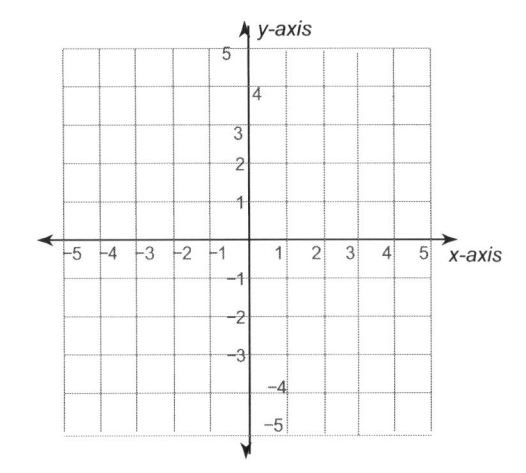

A) $(4, 4)$ B) $(-3, -4)$
C) $(-4, -3)$ D) $(-3, 4)$

X	Y
0	
	0

241

Grade 8

Vol 1
Week 14
System of equations

94) Find the solution of the system of equations given below by graphing

$$y = \frac{3}{4}x + 1$$

$$y = -\frac{1}{4}x - 3$$

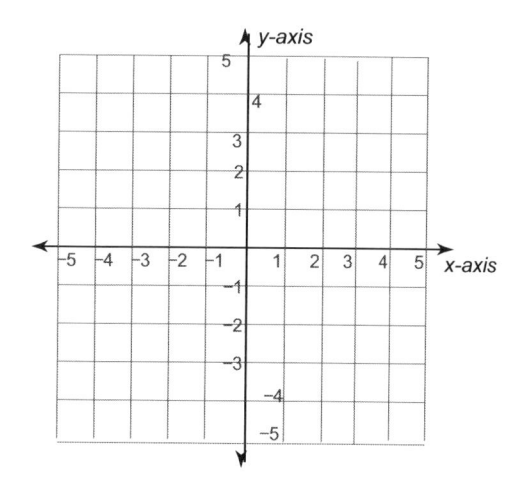

A) $(-2, -4)$ B) $(4, -5)$
C) $(-5, 4)$ D) $(-4, -2)$

X	Y
0	
	0

95) Find the solution of the system of equations given below by graphing

$$y = \frac{1}{2}x + 2$$

$$y = 2x - 1$$

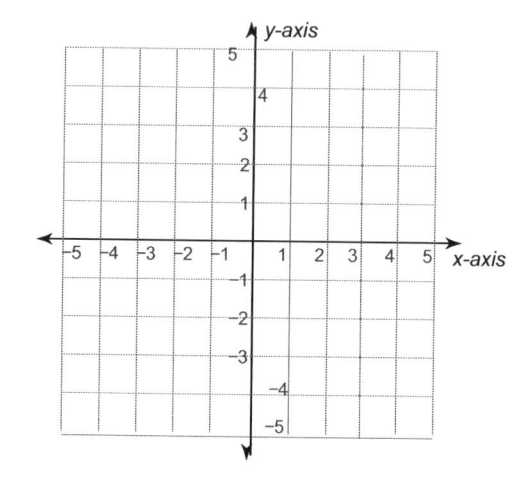

A) $(-3, 2)$ B) $(2, -3)$
C) $(2, 3)$ D) $(-2, -3)$

X	Y
0	
	0

Grade 8

Vol 1 Week 14 System of equations

96) Find the solution of the system of equations given below by graphing

$$y = -\frac{1}{4}x + 3$$

$$y = x - 2$$

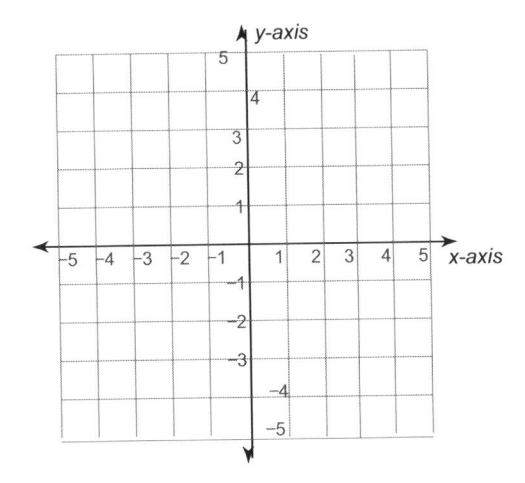

A) $(2, 4)$ B) $(-5, 4)$
C) $(-2, 4)$ D) $(4, 2)$

X	Y
0	
	0

97) Find the solution of the system of equations given below by graphing

$$y = \frac{3}{2}x - 2$$

$$y = \frac{3}{2}x + 2$$

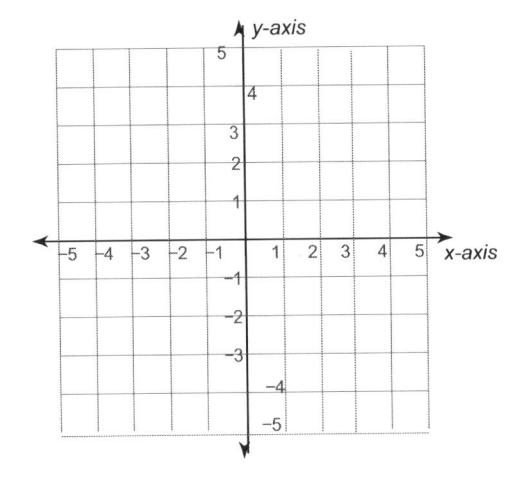

A) $(1, -2)$ B) $(1, 2)$
C) No solution D) $(2, -2)$

X	Y
0	
	0

Grade 8

Vol 1
Week 14
System of equations

98) Find the solution of the system of equations given below by graphing

$$y = \frac{4}{3}x + 2$$

$$y = -\frac{1}{3}x - 3$$

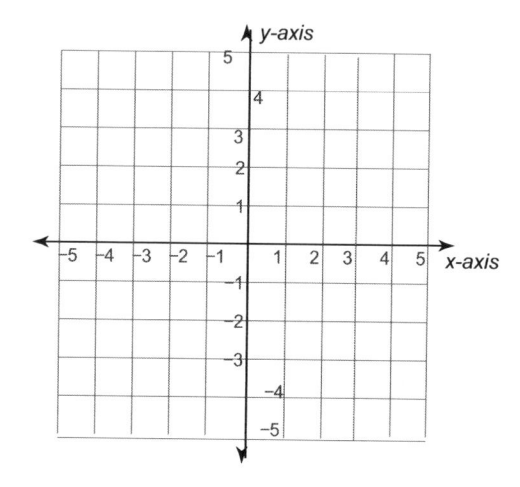

A) $(3, -3)$ B) $(-3, -2)$

C) $(-3, 3)$ D) $(-3, -3)$

X	Y
0	
	0

99) Find the solution of the system of equations given below by graphing

$$x - 3y = 6$$

$$4x - 3y = -3$$

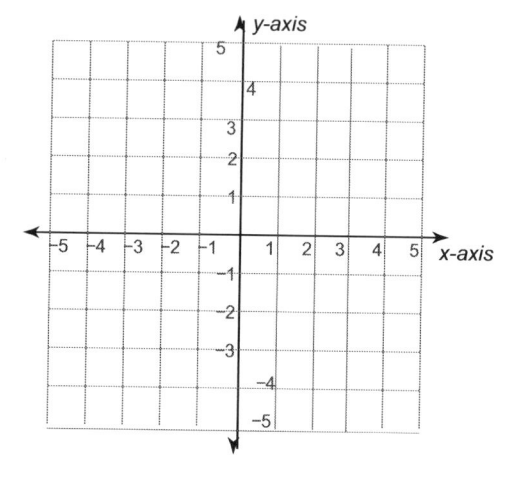

A) $(-3, 3)$ *B) $(-3, -3)$

C) $(3, -3)$ D) $(3, 3)$

X	Y
0	
	0

©All rights reserved-Math-Knots LLC., VA-USA

Grade 8

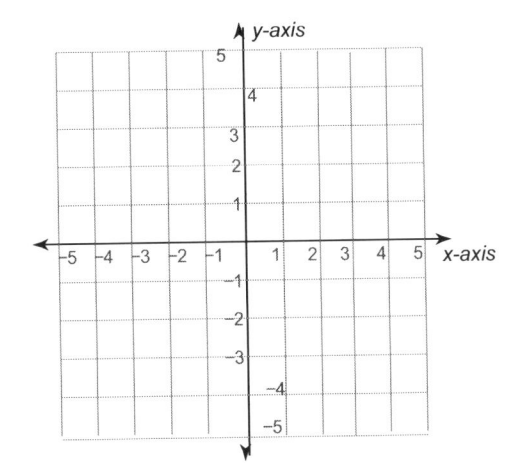

Vol 1 Week 14 System of equations

100) Find the solution of the system of equations given below by graphing

$$2x + y = 2$$
$$2x + y = 1$$

A) No solution

B) $(2, 3)$

C) $(-2, -3)$

D) Infinite number of solutions

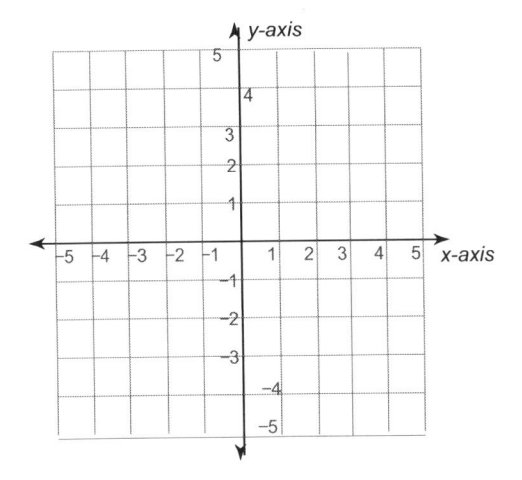

X	Y
0	
	0

101) Find the solution of the system of equations given below by graphing

$$x - 2y = -4$$
$$x - 2y = 8$$

A) $(-1, 2)$ B) $(-1, -4)$

C) $(-1, -1)$ D) No solution

X	Y
0	
	0

Grade 8

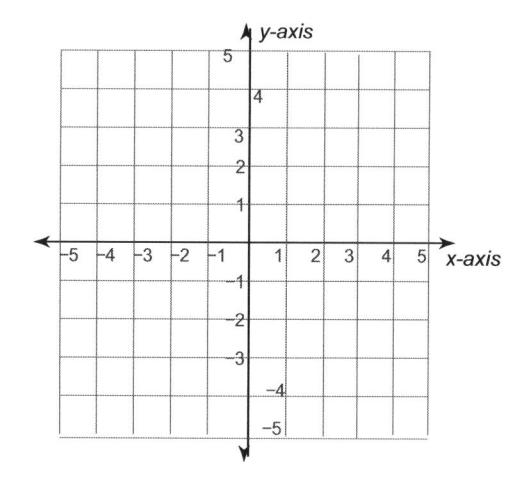

Vol 1 Week 14 System of equations

102) Find the solution of the system of equations given below by graphing

$$3x - 2y = -8$$
$$x + y = -1$$

A) $(-2, 1)$ B) $(1, -2)$

C) $(2, 1)$ D) $(1, 2)$

X	Y
0	
	0

103) Find the solution of the system of equations given below by graphing

$$2x - y = -4$$
$$x - 4y = 12$$

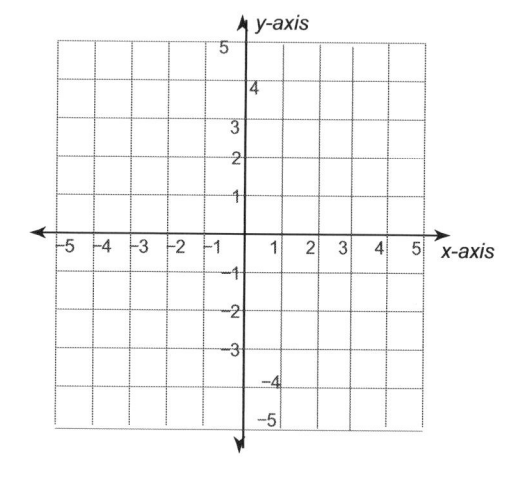

A) $(-4, -4)$ B) $(1, 4)$

C) $(-4, 1)$ D) $(-4, 4)$

X	Y
0	
	0

Grade 8

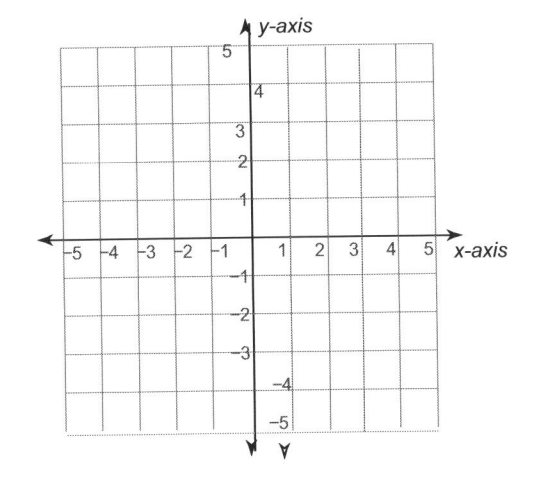

Vol 1 Week 14 — System of equations

104) Find the solution of the system of equations given below by graphing

$$5x + y = -1$$
$$5x + y = 1$$

A) No solution B) $(3, 2)$
C) $(-3, -2)$ D) $(3, -2)$

X	Y
0	
	0

105) Find the solution of the system of equations given below by graphing

$$x + 4y = 16$$
$$7x - 4y = 16$$

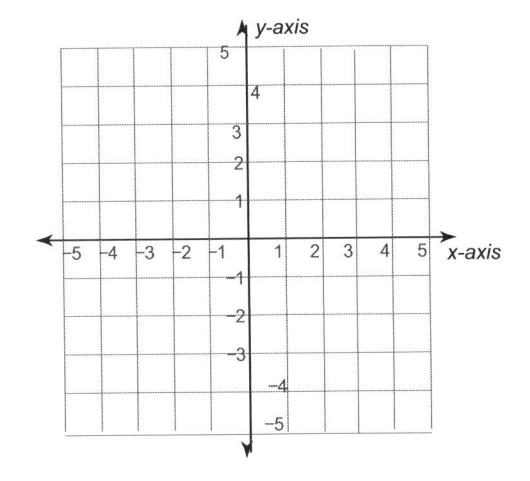

A) $(4, 3)$ B) $(-1, -4)$
C) $(4, 4)$ D) $(4, -4)$

X	Y
0	
	0

Grade 8

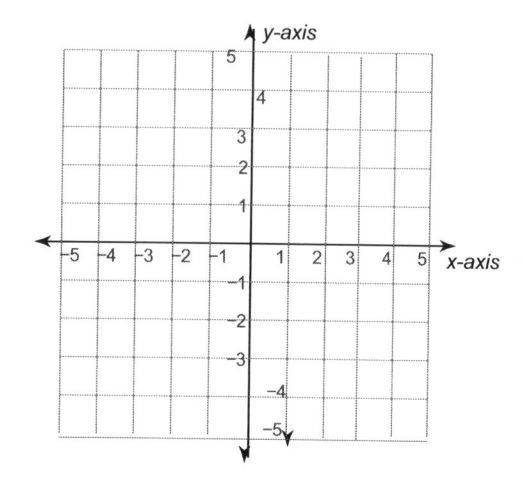

Vol 1 Week 14 System of equations

106) Find the solution of the system of equations given below by graphing

$$x + 2y = 6$$
$$x + 2y = -6$$

A) $(-2, 1)$ B) $(-2, 2)$

C) $(-1, -2)$ D) No solution

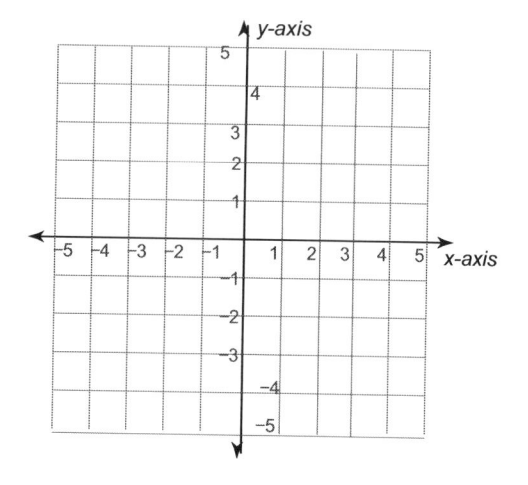

X	Y
0	
	0

107) Find the solution of the system of equations given below by graphing

$$x - 2y = 6$$
$$3x - y = -2$$

A) $(-4, 2)$ B) $(2, -4)$

C) $(-2, 4)$ D) $(-2, -4)$

X	Y
0	
	0

Grade 8

Vol 1 Week 14 System of equations

108) Find the solution of the system of equations given below by graphing

$$5x - y = 4$$
$$2x + y = 3$$

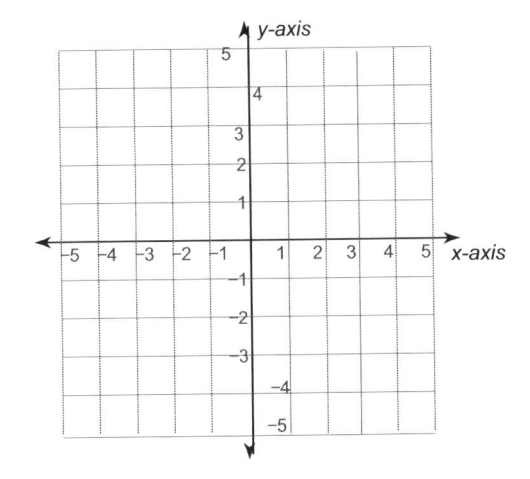

A) (1, 1)   B) (−1, −4)

C) (−1, −1)   D) (−1, 1)

X	Y
0	
	0

109) Find the solution of the system of equations given below by graphing

$$7x + 4y = 16$$
$$7x + 4y = -8$$

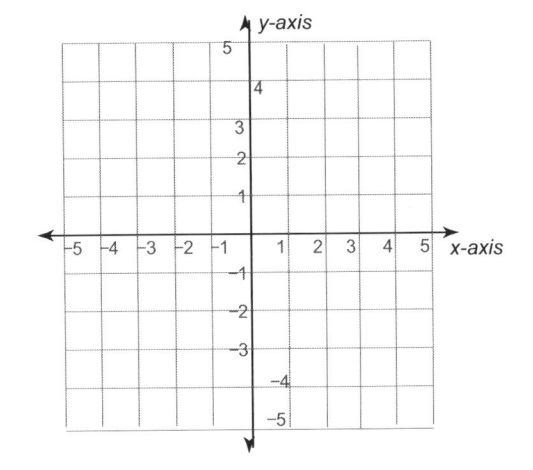

A) (−4, 3)   B) No solution

C) (−4, −3)   D) (−2, −4)

X	Y
0	
	0

Grade 8

Vol 1, Week 14 — System of equations

---

110) Find the solution of the system of equations given below by graphing

$$7x - 3y = 12$$
$$x - 3y = -6$$

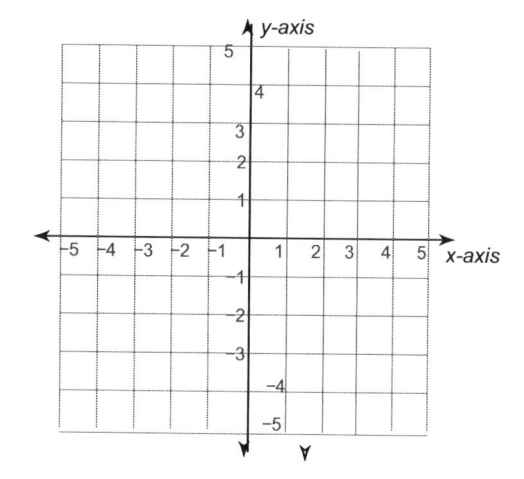

A) No solution

B) Infinite number of solutions

C) $(3, -3)$

D) $(3, 3)$

X	Y
0	
	0

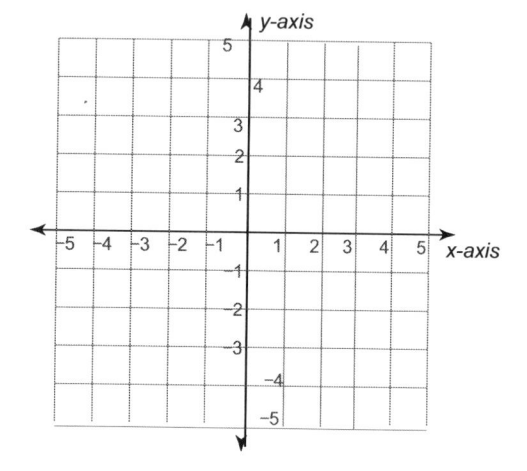

111) Find the solution of the system of equations given below by graphing

$$x + y = -4$$
$$y = -3$$

A) $(5, 3)$ B) $(5, 4)$

C) $(-5, -3)$ D) $(-1, -3)$

X	Y
0	
	0

Grade 8

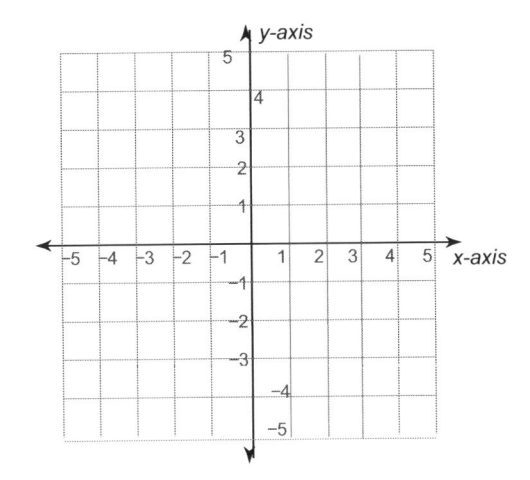

Vol 1 Week 14 System of equations

112) Find the solution of the system of equations given below by graphing

$$8x + y = -4$$
$$x + y = 3$$

A) $(-1, -4)$ B) $(1, 4)$

C) $(-1, 4)$ D) $(1, -4)$

X	Y
0	
	0

113) Find the solution of the system of equations given below by graphing

$$7x - 3y = -12$$
$$7x - 3y = 3$$

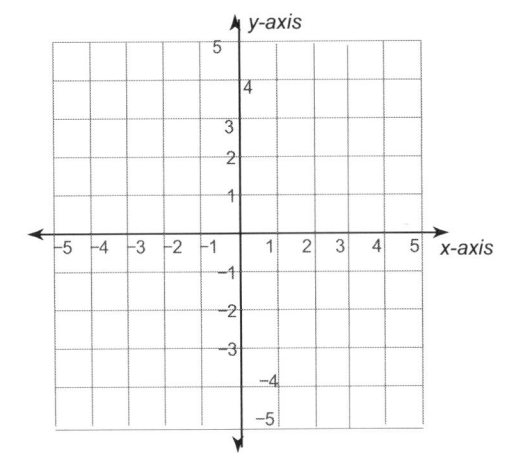

A) $(1, 5)$ B) $(5, 1)$

C) $(-5, 1)$ D) No solution

X	Y
0	
	0

251

Grade 8

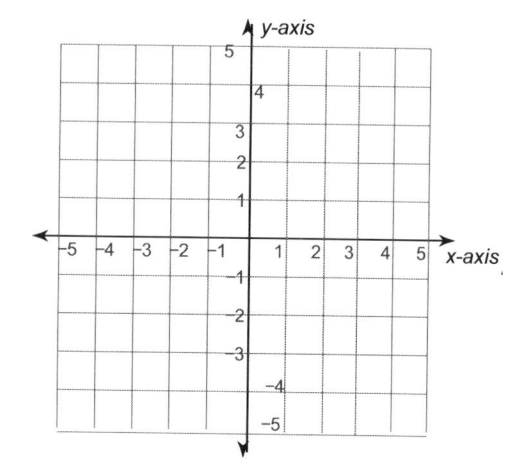

Vol 1 Week 14 System of equations

114) Find the solution of the system of equations given below by graphing

$$x + y = 3$$
$$2x - y = 3$$

A) (3, 5)   B) (2, 1)

C) (2, 5)   D) (3, –5)

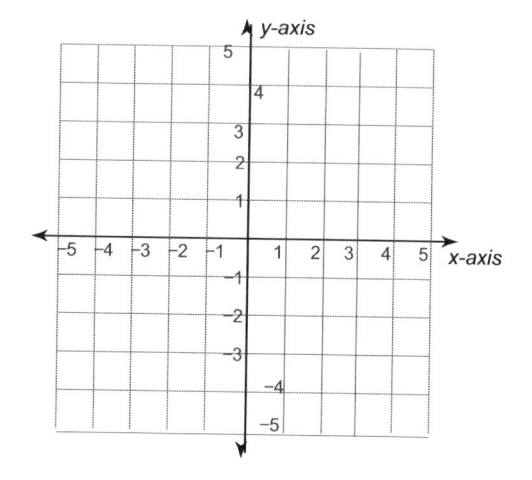

X	Y
0	
	0

115) Find the solution of the system of equations given below by graphing

$$x + 4y = -4$$
$$x - 2y = 8$$

A) (–5, –2)   B) (5, –2)

C) (4, –2)   D) (–2, –4)

X	Y
0	
	0

Grade 8

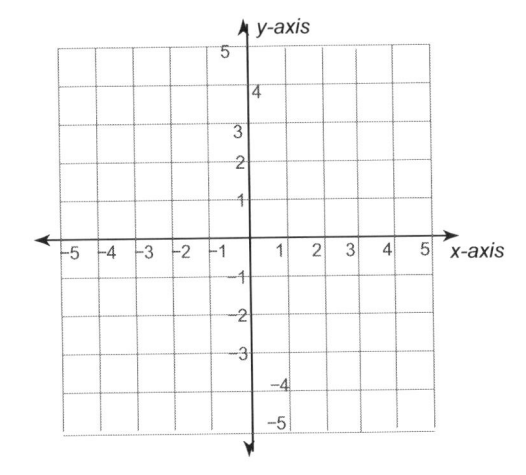

Vol 1
Week 14
System of equations

116) Find the solution of the system of equations given below by graphing

$$5x - y = 2$$
$$x - y = -2$$

A) $(-3, -2)$ B) $(-3, -1)$
C) $(1, 3)$ D) $(-3, -3)$

X	Y
0	
	0

117) Find the solution of the system of equations given below by graphing

$$x + 3y = 12$$
$$4x - 3y = 3$$

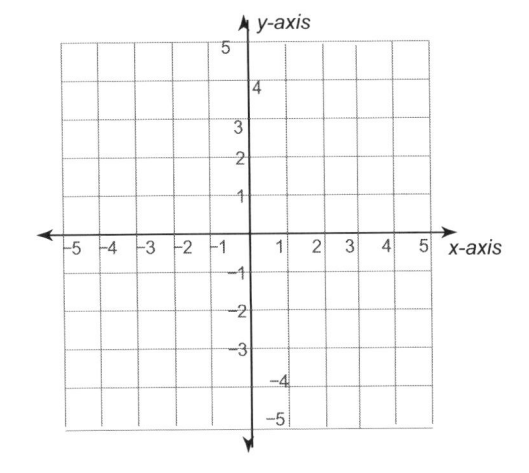

A) $(-3, 3)$ B) $(3, -3)$
C) $(-3, -3)$ D) $(3, 3)$

X	Y
0	
	0

Grade 8

Vol 1 Week 14 System of equations

118) Find the solution of the system of equations given below by graphing

$$2x + y = 4$$
$$x - y = -1$$

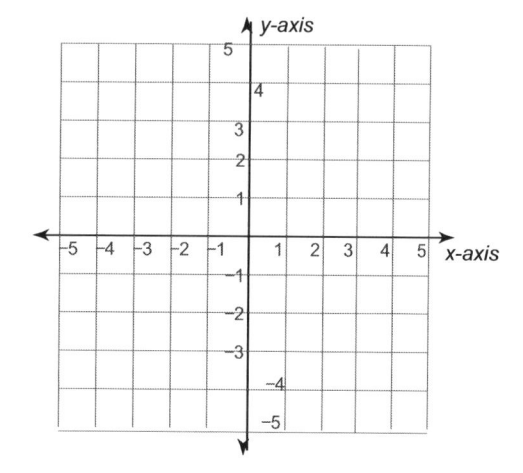

A) $(-5, -1)$ B) $(-1, -5)$
C) $(1, 2)$ D) $(-1, 2)$

X	Y
0	
	0

119) Find the solution of the system of equations given below by graphing

$$2x + y = 4$$
$$x - 3y = 9$$

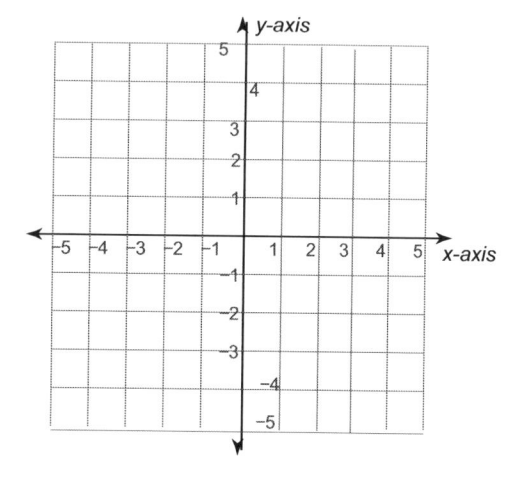

A) $(3, 1)$ B) $(-3, 3)$
C) $(3, -2)$ D) $(-3, -2)$

X	Y
0	
	0

Grade 8

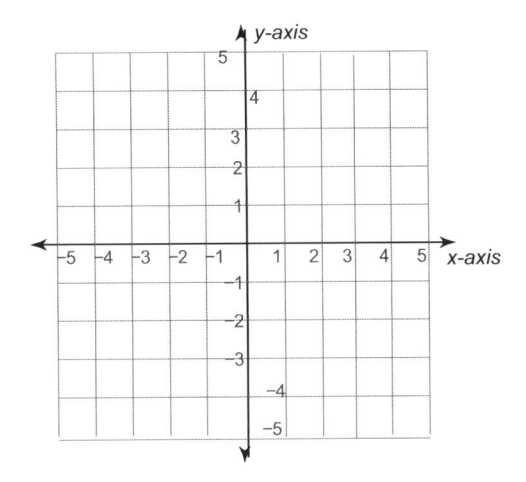

Vol 1 Week 14 System of equations

120) Find the solution of the system of equations given below by graphing

$$x - 2y = -6$$
$$x + y = -3$$

A) $(-3, 1)$ B) No solution
C) $(-4, 1)$ D) $(1, -4)$

X	Y
0	
	0

121) Find the solution of the system of equations given below by graphing

$$3x - 2y = -6$$
$$x - 4y = 8$$

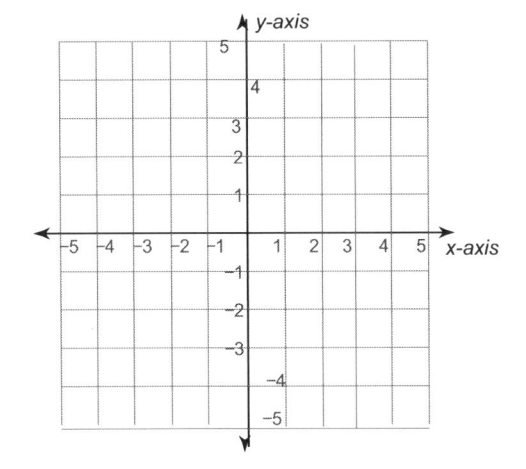

A) $(-4, -3)$ B) $(-3, -4)$
C) $(3, 4)$ D) $(3, -4)$

X	Y
0	
	0

Grade 8

Vol 1 Week 14 System of equations

122) Find the solution of the system of equations given below by graphing

$$3x + 4y = 8$$
$$x - 2y = 6$$

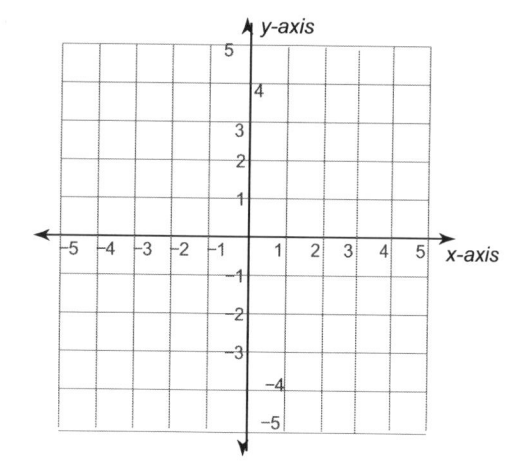

A) $(1, 4)$

B) $(4, 1)$

C) $(4, -1)$

D) Infinite number of solutions

X	Y
0	
	0

Andy bought 7 puzzle sets for a total of $176. puzzle A cost $30 and puzzle B cost $13.

123) How many number of puzzle A's did he buy?

A) 4 puzzle A   B) 5 puzzle A

C) 3 puzzle A   D) 2 puzzle A

124) How many number of puzzle B's did he buy?

A) 5 puzzle B   B) 2 puzzle B

C) 4 puzzle B   D) 3 puzzle B

Kate sold 10 art pieces for a total of $115. Art A cost $8 and Art B cost $15.

125) How many number of Art A pieces did she sold?

A) 5 Art A   B) 7 Art A

C) 8 Art A   D) 3 Art A

126) How many number of Art B pieces did she sold?

A) 3 Art B   B) 2 Art B

C) 5 Art B   D) 4 Art B

Grade 8

Vol 1 Week 14 System of equations

Bryan spent $610 on books. Sports books cost $70 and autobiography books cost $80. He bought a total of 8 books.

127) How many number of sports books did she buy ?

A) 3 sports books B) 2 sports books

C) 6 sports books D) 7 sports books

128) How many number of autobiography books did she buy ?

A) 4 autobiography books

B) 5 autobiography books

C) 2 autobiography books

D) 2 autobiography books

Mary bought 8 plants for a total of $20. Lily plant cost $2 and rose plant cost $3.

129) How many number of Lily plants did she buy ?

A) 4 Lily plant B) 5 Lily plant

C) 6 Lily plant D) 4 Lily plant

Oak middle school took 205 students from grade 6 on a field trip. They were transported in 12 vehicles, in mini vans and buses. Each mini van holds 6 students and each bus hold 25 students.

130) Find the number of mini vans used for transportation.

A) 5 mini vans B) 8 mini vans

C) 10 mini vans D) 9 mini vans

131) Find the number of buses used for transportation.

A) 5 buses B) 4 buses

C) 2 buses D) 7 buses

Prim rose elementary took grade 5 students to a science museum. Students are transported in 8 vehicles of cars and buses together. Each car holds 5 students and each bus holds 55 students. A total of 190 students visited the science museum.

132) Find the number of cars used for transportation ?

A) 6 cars B) 6 cars

C) 7 cars D) 5 cars

133) Find the number of cars used for transportation ?

A) 4 buses B) 2 buses

C) 4 buses D) 3 buses

Grade 8

Vol 1 Week 14 System of equations

220 students from STEM Club participated in a competition this weekend. Few students came in cars in groups of 5 students and other students took buses in groups of 50 students. A total of 8 vehicles were used by students.

134) How many cars were used for transportation ?

A) 4 cars   B) 6 cars

C) 2 cars   D) 7 cars

135) How many buses were used for transportation ?

A) 3 buses   B) 2 buses

C) 4 buses   D) 5 buses

A kindergarten class of 111 students went on a field trip to a pumpkin farm in cars and mini vans. A total of 11 vehicles are used for transportation. Each car carried 5 students and each mini van carried 13 students.

136) How many cars were used for field trip ?

A) 5 cars   B) 6 cars

C) 7 cars   D) 4 cars

137) How many mini vans were used for field trip ?

A) 3 mini vans   B) 2 mini vans

C) 7 mini vans   D) 4 mini vans

A class used mini buses and buses to go on a field trip. They used 8 vehicles to go on the trip. Each van holds 10 students and each bus holds 50 students. 280 students went on the trip

138) How many number of mini buses did the class use ?

A) 8 mini buses   B) 3 mini buses

C) 6 mini buses   D) 4 mini buses

139) How many number of buses did the class use ?

A) 3 buses   B) 4 buses

C) 2 buses   D) 5 buses

Robotics team of 130 students went to compete in nationals at Washington DC. Students were transported in cars and mini vans. Each car has a capacity of 5 students and each mini van has a capacity of 15 students. A total of 10 vehicles were used for transportation.

140) How many cars were used by Robotics team for transportation?

A) 6 cars   B) 5 cars

C) 8 cars   D) 2 cars

141) How many mini vans were used by Robotics team for transportation?

A) 8 mini vans   B) 5 mini vans

C) 2 mini vans   D) 3 mini vans

Grade 8

**Vol 1
Week 14
System of equations**

A grade 7 class of 112 students are participating at a county music festival representing their school. Students were transported in 16 vehicles. A group of 5 students traveled in mini cars and groups of nine students traveled in mini vans.

142) How many of mini cars did the class use ?

A) 7 mini cars B) 14 mini cars

C) 8 mini cars D) 16 mini cars

143) How many of mini vans did the class use ?

A) 9 mini vans B) 8 mini vans

C) 4 mini vans D) 3 mini vans

A 200 students of Math Club participated in a math bowl competition. Students car pooled in groups of 5 students in cars and groups of 45 students in mini buses. A total of 16 vehicles were used for transportation.

144) How many of cars were used ?

A) 5 cars B) 13 cars

C) 17 cars D) 14 cars

Grade 8

**Vol 1
Week 15
Assessment 2**

1) Find the slope of the line from the graph.

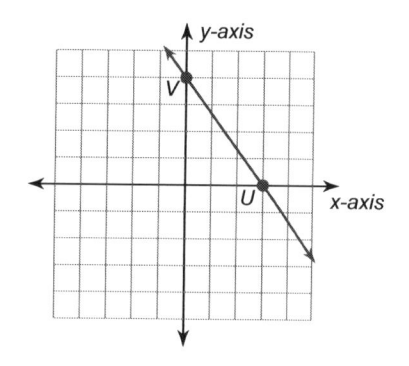

A) $\dfrac{3}{4}$ B) $-\dfrac{4}{3}$

C) $-\dfrac{3}{4}$ D) $\dfrac{4}{3}$

2) Find the slope of the line from the graph.

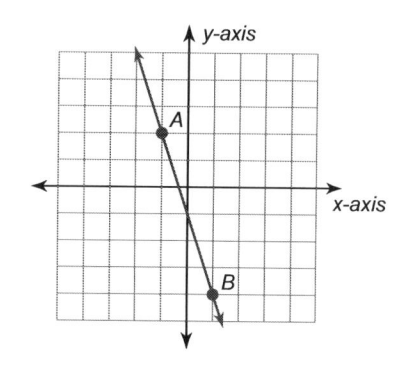

A) $\dfrac{1}{3}$ B) -3

C) 3 D) $-\dfrac{1}{3}$

3) Find the slope of the line from the graph.

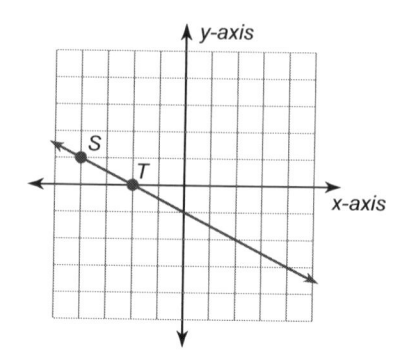

A) $-\dfrac{1}{2}$ B) 2

C) -2 D) $\dfrac{1}{2}$

4) Find the slope of the line passing through the points P (19, 16) and Q (-6, 16)

A) 1   B) Undefined

C) −1   D) 0

5) Find the slope of the line passing through the points P (18, 7) and Q (-12, -4)

A) $\dfrac{30}{11}$ B) $-\dfrac{30}{11}$

C) $\dfrac{11}{30}$ D) $-\dfrac{11}{30}$

©All rights reserved-Math-Knots LLC., VA-USA   www.math-knots.com | www.a4ace.com

Grade 8

Vol 1
Week 15
Assessment 2

6) Which of the below straight line has the least slope?

A) $y = x - \dfrac{1}{3}$ B) $y = -\dfrac{1}{3}x - \dfrac{1}{3}$

C) $y = \dfrac{1}{3}x - \dfrac{1}{3}$ D) $y = -3x + 1$

7) Find the value of b that satisfies the below equation

$$5(3b - 4) = -125$$

A) 8   B) -7

C) 10   D) -13

8) Find the missing coordinate given the points on the straight line with a slope of $-\dfrac{1}{9}$ passing through the points

L(x,-12) and M(13,-14)

A) -5   B) -13

C) -11   D) -1

9) Find the value of x that satisfies the below equation

$$-5(x - 4) - 5x = 8(-1 - 3x)$$

A) -2   B) 14

C) -14   D) No solution.

10) Find the missing coordinate given the points on the straight line with a slope of $-\dfrac{9}{7}$ passing through the points

S(-4,y) and T(-11,11)

A) 2   B) 8

C) 0   D) 6

11) Find the slope of the straight line given below.

$$y = -\dfrac{8}{3}x - 5$$

A) $\dfrac{3}{8}$ B) $-\dfrac{8}{3}$

C) $\dfrac{8}{3}$ D) $-\dfrac{3}{8}$

12) Which of the below straight line has the highest slope?

A) $-y = 0$ B) $y = 2x + 5$

C) $9x + 7 = y$ D) $y = -2x + 5$

13) Find the distance between the points P(-12,9) and Q(2,-3)

A) $2\sqrt{85}$ B) 8

C) $\sqrt{26}$ D) $2\sqrt{34}$

©All rights reserved-Math-Knots LLC., VA-USA   261   www.math-knots.com | www.a4ace.com

Grade 8

**Vol 1
Week 15
Assessment 2**

14) Which of the below straight line has the highest slope?

   A) $x = 5$ B) $y = x$
   C) $x = -1$ D) $y = 1$

15) Which of the below straight line has the least slope?

   A) $y = \frac{4}{5}x + \frac{1}{5}$ B) $y = \frac{16}{5}x + \frac{1}{5}$
   C) $y = \frac{1}{5}x - \frac{16}{5}$ D) $y = -\frac{16}{5}x + \frac{1}{5}$

16) Write the standard form of the equation of the straight line passing through P(4,-1) and parallel to $y = x + 2$

   A) $x - y = 5$ B) $x + y = 5$
   C) $x + y = -5$ D) $4x - y = 5$

17) Find the slope of the straight line given below.

   $$y = -\frac{1}{2}x$$

   A) $-\frac{1}{2}$ B) 2
   C) $\frac{1}{2}$ D) -2

18) Find the slope of the straight line given below.

   $$y = -2x - 4$$

   A) 2 B) $-\frac{1}{2}$
   C) -2 D) $\frac{1}{2}$

19) Find the slope of the straight line given below.

   $$y = 7x + 3$$

   A) 7 B) $-\frac{1}{7}$
   C) -7 D) $\frac{1}{7}$

20) Which of the below straight line has the highest slope?

   A) $8x - 6y = -1$
   B) $8x + 6y = -1$
   C) $x - 6y = -8$
   D) $8x + 6y = 1$

21) Which of the below straight line has the least slope?

   A) $3x + y = 15$
   B) $2x - 3y = -1$
   C) $x + 12y = 3$
   D) $x - 3y = -12$

©All rights reserved-Math-Knots LLC., VA-USA   www.math-knots.com | www.a4ace.com

Grade 8

**Vol 1
Week 15
Assessment 2**

22) Find the equation of the straight line passing through P(3,5) and parallel to y = -2x - 1

A) y = 2x + 11   B) y = -2x + 11

C) y = -5x + 11   D) y = 3x + 11

23) Find the equation of the straight line passing through P(1,-5) and parallel to y = -7x - 2

A) y = 2x - 1   B) y = -7x + 2

C) y = x + 2   D) y = -x + 2

24) Write the standard form of the equation of the straight line passing through P(-1,-2) and parallel to y = 3x - 4

A) 3x - y = -1   B) 3x - y = 4

C) x - 2y = -10   D) 3x + 4y = -1

25) Find the equation of the straight line passing through P(4,1) and parallel to $y = \frac{3}{4}x - 1$

A) $y = -\frac{3}{4}x - 2$ B) $y = \frac{5}{4}x - \frac{3}{4}$

C) $y = \frac{3}{4}x - 2$ D) $y = -2x - \frac{3}{4}$

26) Write the standard form of the equation of the straight line passing through P(-2,-3) and parallel to y = 3x + 2

A) 3x - y = -3   B) 3x - y = 3

C) 5x + y = 3   D) 4x - y = -1

27) Find the midpoint of the points P(-20,-7) and Q(-9,-3)

28) Find the value of x that satisfies the below equation
$-4(n + 7) = -8(n + 2)$

A) -13   B) 3

C) 6   D) -6

Grade 8

29) Write the slope intercept form of the equation of the straight line passing through P(4,3) and perpendicular to y = -2x + 3

A) $y = x + \dfrac{1}{2}$ B) $y = -x + \dfrac{1}{2}$

C) $y = -\dfrac{5}{2}x + \dfrac{1}{2}$ D) $y = \dfrac{1}{2}x + 1$

30) Write the slope intercept form of the equation of the straight line passing through P(0,2) and perpendicular to y = x + 3

A) $y = 2x - 1$ B) $y = 2x + 2$

C) $y = -2x + 2$ D) $y = -x + 2$

31) Write the standard form of the equation of the straight line passing through P(-2,-5) and perpendicular to $y = -\dfrac{7}{9}x - 3$

A) $9x - 17y = -17$

B) $9x - 7y = 17$

C) $17x - 9y = -17$

D) $9x + 17y = -7$

32) Write the slope intercept form of the equation of the straight line passing through P(4,-5) and perpendicular to $y = \dfrac{1}{2}x - 5$

A) $y = -2x + 3$ B) $y = 5x + 3$

C) $y = -4x + 3$ D) $y = 4x + 3$

33) Write the standard form of the equation of the straight line passing through P(5,4) and perpendicular to y = -5x - 2

A) $2x + 5y = 15$

B) $x - 5y = -15$

C) $x + 15y = 5$

D) $3x - 5y = 15$

34) Write the standard form of the equation of the straight line passing through P(-3,-4) and perpendicular to y = -x + 5

A) $x - 2y = -1$ B) $x - y = -1$

C) $x - y = 1$ D) $x + y = -1$

Grade 8

Vol 1 — Week 15 — Assessment 2

35) Find the value of n that satisfies the below equation

$-108 = 4(8n + 5)$

A) −4 B) −3
C) −9 D) No solution.

36) Find the value of p that satisfies the below equation

$5(1 - 7p) = 145$

A) 15 B) 12
C) 3 D) −4

37) Find the value of p that satisfies the below equation

$-8 + 4(1 + 6p) = 188$

A) No solution. B) 2
C) 8 D) 12

38) Find the value of a that satisfies the below equation

$132 = 5(1 - 3a) + 7$

A) −13 B) −8
C) All real numbers D) 10

39) Find the value of n that satisfies the below equation

$85 = 5(5 + 2n)$

A) −14 B) 12
C) 8 D) 6

40) Find the value of r that satisfies the below equation

$-2(7r + 5) = -94$

A) All real numbers B) 12
C) −7 D) 6

41) Find the value of a that satisfies the below equation

$-4(2 + 3a) = -92$

A) 2 B) −1
C) 7 D) 1

42) Find the value of k that satisfies the below equation

$-4(1 + 8k) = 220$

A) −7 B) −1
C) 10 D) 11

Grade 8

Vol 1
Week 15
Assessment 2

43) Find the value of b that satisfies the below equation
$$-13 = 5(7b - 2) - (3 - b)$$
A) −7   B) −12
C) 0   D) 6

44) Find the value of a that satisfies the below equation
$$-3(-4 - 5a) - 5(2 + 7a) = 2$$
A) 5   B) 14
C) 0   D) −14

45) Find the value of x that satisfies the below equation
$$2(7 + 4x) + 3(4 + 4x) = -14$$
A) −2   B) All real numbers
C) −4   D) 2

46) Find the value of n that satisfies the below equation
$$-33 = 7(6n + 3) + 4(n - 2)$$
A) −14   B) 11
C) −1   D) No solution.

47) Find the value of r that satisfies the below equation
$$-6(6r + 2) - 5(r - 6) = -64$$
A) 2   B) 3
C) 0   D) 4

48) Find the endpoint of the line segment given one endpoint as (13,17) and midpoint as Q(-35,35)
A) (−11, 26)   B) (3, −13)
C) (61, −1)   D) (−83, 53)

49) Find the endpoint of the line segment given one endpoint as (-18,5) and midpoint as Q(-37,9)
A) (−23, −5)   B) (−27.5, 7)
C) (−56, 13)   D) (1, 1)

50) Find the midpoint of the points P(-2,-5) and Q(5,-7)

51) Find the midpoint of the points P(-14,1) and Q(-9,4)

52) Find the value of r that satisfies the below equation
$$-432 = -8(5 + 7r)$$
A) −7   B) 7
C) 15   D) 8

Grade 8

Vol 1
Week 15
Assessment 2

53) Find the midpoint of the line segment plotted below

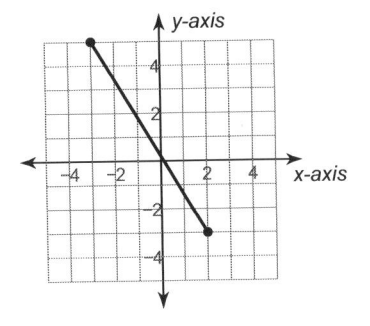

54) Find the midpoint of the points P(-9,-18) and Q(2,2)

55) Find the value of x that satisfies the below equation

$-6(4m+6) = -5(7m-3) + 4$

A) 13   B) No solution.

C) 5   D) 9

56) Find the value of x that satisfies the below equation

$6 + 6(x+6) = 6(3x+6) - 6x$

A) 0   B) All real numbers

C) 1   D) 15

57) Find the value of x that satisfies the below equation

$-2x + 6(x-1) = 3(x+3)$

A) 15   B) 14

C) -8   D) 4

58) Find the value of x that satisfies the below equation

$$-\frac{19141}{216} = \frac{15}{4}x + \frac{14}{3}\left(\frac{8}{3}x - 8\right)$$

A) $-\dfrac{19}{6}$ B) $-2\dfrac{3}{16}$

C) $-\dfrac{2}{3}$ D) $\dfrac{8}{13}$

59) Find the value of x that satisfies the below equation

$$-\frac{171}{2} = \frac{15}{4}\left(-\frac{28}{5}k + 1\right)$$

A) $\dfrac{8}{13}$ B) $-\dfrac{3}{7}$

C) $\dfrac{6}{7}$ D) $\dfrac{17}{4}$

60) Find the missing coordinate given the points on the straight line with a slope of 11 passing through the points M(x , -8) and N(5 , 14)

A) 3   B) 11

C) -13   D) 10

Grade 8

Vol 1 Week 15 Assessment 2

61) For a field trip 13 students rode in cars and the rest filled three buses. How many students were in each bus if 91 students were on the trip?

   A) 26 B) 27
   C) 30.33 D) 25

62) The sum of three consecutive odd numbers is 75. What is the smallest of these numbers?

   A) 25 B) 27
   C) 21 D) 23

63) You had $21 to spend on seven pens. After buying them you had $0. How much did each pen cost?

   A) $0 B) $3
   C) $4 D) $2

64) You bought a magazine for $5 and some erasers for $2 each. You spent a total of $21. How many erasers did you buy?

   A) 7 B) 4
   C) 6 D) 8

65) Find the distance between the points P(6,4) and Q(0,8)

   A) $6\sqrt{3}$ B) $\sqrt{10}$
   C) $2\sqrt{13}$ D) $6\sqrt{5}$

66) Amanda's Bikes rents bikes for $17 plus $4 per hour. Wilbur paid $29 to rent a bike. For how many hours did he rent the bike?

   A) 2 B) 1
   C) 7.25 D) 3

67) Find the distance between the points P(-2.2,-0.7) and Q(-3.1,7.3). (Round to the nearest tenth)

   A) 8.1 B) 3
   C) 3.4 D) 8.5

68) Find the distance between the points P(-1.3,-2.6) and Q(-2.5,-6.6). (Round to the nearest tenth)

   A) 2.3 B) 9.9
   C) 10 D) 4.2

69) Find the distance between the points P(-3.3,-7.9) and Q(-8,1.8). (Round to the nearest tenth)

   A) 12.8 B) 3.8
   C) 10.8 D) 9.5

70) Find the distance between the points P(-8,-11) and Q(-1,-7)

   A) $9\sqrt{3}$ B) 3
   C) $\sqrt{11}$ D) $\sqrt{65}$

Grade 8

Vol 1 — Week 15 — Assessment 2

71) Find the distance between the points plotted below

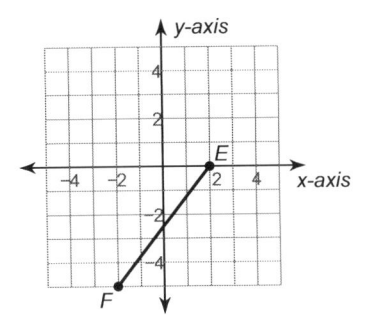

A) 5 B) $\sqrt{5}$

C) $\sqrt{41}$ D) 3

72) Find the distance between the points plotted below

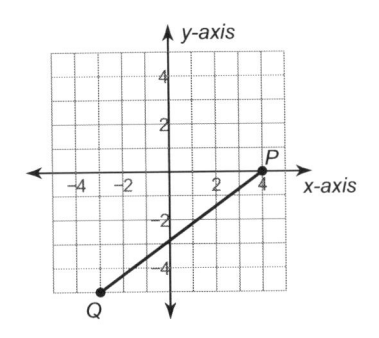

A) $\sqrt{26}$ B) $\sqrt{6}$

C) $2\sqrt{3}$ D) $\sqrt{74}$

73) Find the value of x that satisfies the below equation

$$8\left(-\frac{23}{6}x + \frac{11}{7}\right) = \frac{824}{7}$$

A) $-15\frac{3}{13}$ B) $-\frac{24}{7}$

C) $\frac{5}{16}$ D) $2\frac{1}{4}$

74) Find the missing coordinate given the points on the straight line with a slope of 0 passing through the points Q(-7,-1) and R(-9,y)

A) 0 B) 4

C) -1 D) -2

75) Find the value of x that satisfies the below equation

$$-\frac{403}{3} = \frac{2}{3} + 5\left(\frac{7}{2}a + 1\right)$$

A) $7\frac{1}{14}$ B) $\frac{23}{15}$

C) -8 D) $\frac{3}{2}$

Grade 8

Vol 1
Week 15
Assessment 2

76) Find the midpoint of the line segment plotted below

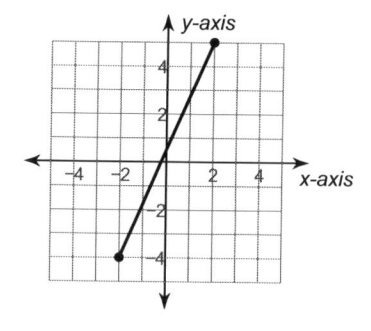

77) Find the distance between the points plotted below

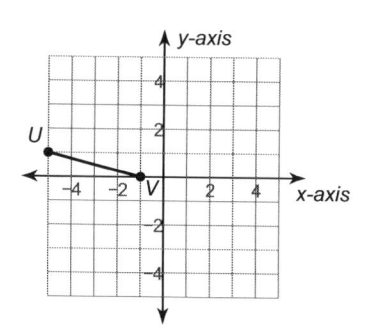

A) $\sqrt{5}$ B) $\sqrt{7}$

C) $\sqrt{37}$ D) $\sqrt{17}$

78) Find the solution of the system of equations given below by graphing

$9 + 3y = 3x$

$3 + 3y + 3x = 0$

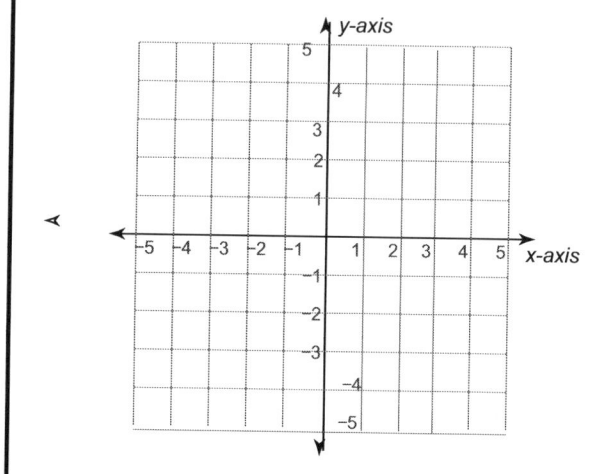

A) $(1, -2)$ B) $(-2, -1)$

C) $(-4, -1)$ D) $(-2, 1)$

X	Y
0	
	0

Grade 8

Vol 1
Week 15
Assessment 2

79) Solve the below system of equations by using the method of elimination

$15x + 5y = -9$
$-9x - 3y = 12$

A) $(2, -9)$ B) $(-2, -9)$

C) $(-9, 2)$ D) No solution

80) Solve the below system of equations by using the method of elimination

$7x + 7y = -35$
$-8x + 11y = -17$

A) $(-2, 3)$ B) $(3, -2)$

C) $(2, -3)$ D) $(-2, -3)$

81) Solve the below system of equations by using the method of elimination

$14x + 10y = -34$
$11x + 9y = -29$

A) $(10, 2)$ B) $(-1, -2)$

C) $(-1, 2)$ D) $(-9, 2)$

82) Solve the below system of equations by using the method of elimination

$13x + 6y = 42$
$3x + 7y = -24$

A) $(-6, -5)$ B) $(6, -6)$

C) $(-2, -5)$ D) $(6, -5)$

83) Solve the below system of equations by using the method of elimination

$-8x + 2y = 0$
$7x - 7y = -42$

A) $(2, -8)$
B) $(2, 8)$
C) Infinite number of solutions
D) $(-2, -8)$

84) Solve the below system of equations by using the method of elimination

$-9x + 5y = -41$
$6x + 3y = -36$

A) $(-1, 10)$
B) $(-8, -10)$
C) Infinite number of solutions
D) $(-1, -10)$

85) Solve the below system of equations by using the method of elimination

$9x - 3y = 42$
$-10x + 13y = -8$

A) $(5, -4)$ B) $(4, 6)$

C) $(6, 4)$ D) $(-5, -4)$

86) Solve the below system of equations by using the method of elimination

$5x - 5y = 15$
$-4x - 6y = -32$

A) $(2, -10)$ B) $(2, 10)$

C) $(10, 2)$ D) $(5, 2)$

Grade 8

Vol 1
Week 15
Assessment 2

87) Solve the below system of equations by using the method of substitution

$6x - 9y = -30$
$-x + 5y = 5$

A) $(4, 0)$ B) $(-4, 0)$

C) $(-5, 0)$ D) $(11, 0)$

88) Solve the below system of equations by using the method of substitution

$y = -2$
$5x + 9y = 17$

A) Infinite number of solutions
B) $(7, 9)$
C) $(-2, 7)$
D) $(7, -2)$

89) Solve the below system of equations by using the method of substitution

$-8x - 4y = -20$
$y = -3$

A) $(4, -3)$ B) $(3, 8)$

C) $(-4, -3)$ D) $(-4, 3)$

90) Solve the below system of equations by using the method of substitution

$12x - y = 19$
$-2x - 4y = 26$

A) $(-1, 7)$ B) $(1, 7)$

C) $(1, -7)$ D) $(-1, -7)$

91) Solve the below system of equations by using the method of substitution

$8x - 10y = -34$
$-3x + 8y = 17$

A) $(1, 10)$ B) $(-3, 1)$

C) $(1, -3)$ D) $(10, 1)$

92) Solve the below system of equations by using the method of substitution

$-9x + 4y = 20$
$5x - 3y = -15$

A) $(0, 5)$ B) $(-9, -5)$

C) $(5, 5)$ D) $(0, -5)$

93) Solve the below system of equations by using the method of substitution

$-2x + 4y = 16$
$6x - y = -4$

A) $(-4, 0)$
B) $(0, 4)$
C) $(4, 0)$
D) Infinite number of solutions

94) Solve the below system of equations by using the method of substitution

$8x + 2y = 20$
$-24x - 6y = -60$

A) $(-3, 10)$
B) Infinite number of solutions
C) $(-3, -3)$
D) $(-3, -6)$

Grade 8

Vol 1
Week 15
Assessment 2

95) Find the solution of the system of equations given below by graphing

$2y = -5x - 8$

$x = -4 + 2y$

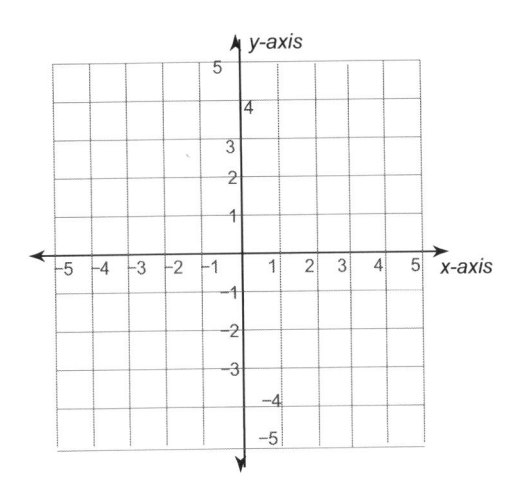

A) No solution B) $(2, -1)$
C) $(-2, -1)$ D) $(-2, 1)$

X	Y
0	
	0

96) Find the solution of the system of equations given below by graphing

$4y = -x + 8$

$15x = -12y - 24$

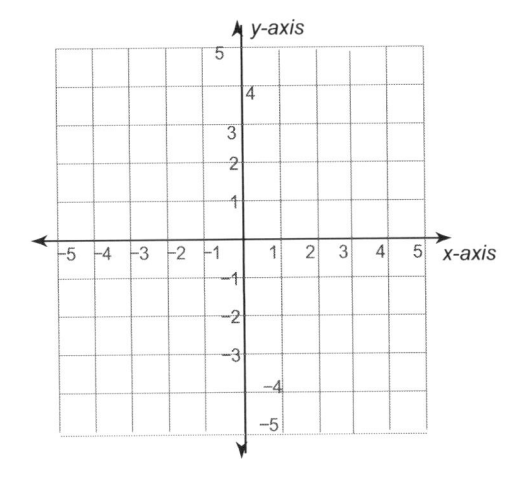

A) $(2, -3)$ B) $(-4, 3)$
C) $(4, -3)$ D) $(4, 3)$

X	Y
0	
	0

273

Grade 8

Vol 1
Week 15
Assessment 2

97) Find the solution of the system of equations given below by graphing

$6 = -3y + x$

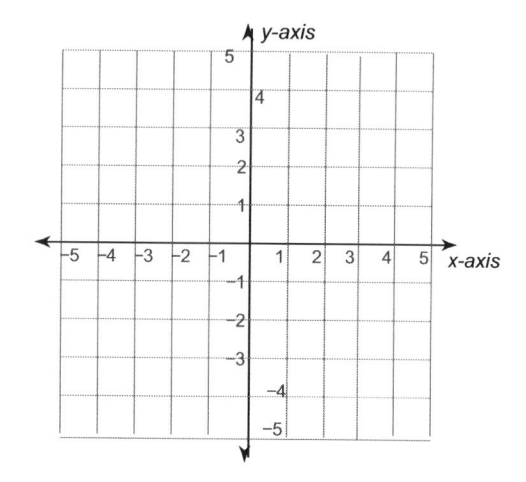

$x + \dfrac{3}{4}y - \dfrac{9}{4} = 0$

A) $(3, -1)$ B) $(3, 1)$
C) $(-1, 3)$ D) $(1, 3)$

X	Y
0	
	0

98) Find the solution of the system of equations given below by graphing

$0 = -y + 2 - x$

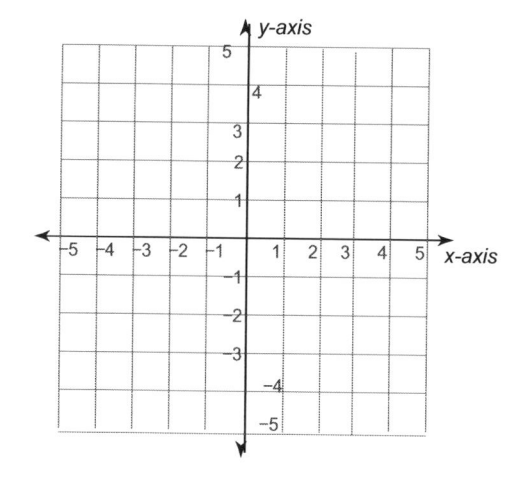

$-1 - 2x = \dfrac{1}{3}y$

A) $(-1, 3)$ B) $(-3, -1)$
C) $(-1, -3)$ D) $(-1, -1)$

X	Y
0	
	0

©All rights reserved-Math-Knots LLC., VA-USA

Grade 8

Vol 1
Week 15
Assessment 2

99) The sum of the digits of a certain two-digit number is 8. When you reverse its digits you decrease the number by 54. What is the number?

100) When you reverse the digits in a certain two-digit number you increase its value by 18. What is the number if the sum of its digits is 6?

101) When you reverse the digits in a certain two-digit number you decrease its value by 18. What is the number if the sum of its digits is 16?

102) Find the value of two numbers if their sum is 21 and their difference is 5.

A) 8 and 13   B) 7 and 14

C) 3 and 7   D) 13 and 20

103) The difference of two numbers is 1. Their sum is 21. What are the numbers?

A) 10 and 11   B) 14 and 16

C) 11 and 13   D) 15 and 15

104) The sum of two numbers is 22. Their difference is 4. Find the numbers.

A) 12 and 14   B) 8 and 20

C) 9 and 13   D) 7 and 10

105) The difference of two numbers is 4. Their sum is 18. Find the numbers.

A) 9 and 7   B) 5 and 14

C) 8 and 9   D) 7 and 11

106) The difference of two numbers is 1. Their sum is 17. Find the numbers.

A) 7 and 4   B) 3 and 8

C) 7 and 12   D) 8 and 9

Lola's school is selling tickets for the annual talent show. On Monday the school sold 7 adult tickets and 14 child tickets for a total of $154. On Tuesday they sold 3 adult tickets and 7 child tickets for $73.

107) Find the price of a adult ticket

108) Find the price of a child ticket.

Grade 8 Answer Keys

Vol 1 Answer Key

Grade 8 — Vol 1 Answer Key

Grade 8

Vol 1 Answer Key

Week 1

#	Ans
1.	A
2.	D
3.	A
4.	C
5.	A
6.	B
7.	D
8.	B
9.	D
10.	A
11.	C
12.	B
13.	D
14.	D
15.	C
16.	C
17.	B
18.	D
19.	B
20.	D
21.	A
22.	C
23.	B
24.	A
25.	D
26.	D
27.	C
28.	A
29.	D

Week 1

#	Ans
30.	B
31.	B
32.	D
33.	D
34.	D
35.	A
36.	C
37.	C
38.	B
39.	C
40.	B
41.	D
42.	B
43.	C
44.	D
45.	B
46.	C
47.	C
48.	C
49.	B
50.	A
51.	D
52.	D
53.	A
54.	D
55.	D
56.	B
57.	D
58.	C

Week 1

#	Ans
59.	B
60.	B
61.	B
62.	C
63.	A
64.	D
65.	D
66.	D
67.	A
68.	C
69.	B
70.	D
71.	B
72.	A
73.	D
74.	C
75.	D
76.	A
77.	A
78.	D
79.	C
80.	C
81.	A
82.	D
83.	B
84.	A
85.	C
86.	A
87.	D

Week 1

#	Ans
88.	C
89.	D
90.	B
91.	A
92.	C
93.	B
94.	D
95.	C
96.	A
97.	D
98.	A
99.	A
100.	C
101.	D
102.	C
103.	D
104.	B
105.	C
106.	A
107.	C
108.	B
109.	A
110.	A

©All rights reserved-Math-Knots LLC., VA-USA

Grade 8

Vol 1 Answer Key

Week 2		Week 2		Week 2		Week 2	
1.	D	30.	B	59.	C	88.	C
2.	B	31.	B	60.	C	89.	D
3.	A	32.	A	61.	B	90.	A
4.	D	33.	A	62.	B	91.	C
5.	A	34.	B	63.	D	92.	C
6.	B	35.	A	64.	A	93.	C
7.	A	36.	A	65.	C	94.	C
8.	C	37.	B	66.	B	95.	A
9.	A	38.	D	67.	B	96.	B
10.	A	39.	C	68.	C	97.	C
11.	D	40.	D	69.	C	98.	A
12.	A	41.	B	70.	D	99.	D
13.	A	42.	D	71.	A	100.	B
14.	B	43.	D	72.	A	101.	C
15.	B	44.	A	73.	A	102.	D
16.	A	45.	B	74.	B	103.	D
17.	C	46.	A	75.	C	104.	C
18.	C	47.	A	76.	C	105.	A
19.	C	48.	D	77.	B	106.	A
20.	B	49.	A	78.	C	107.	B
21.	C	50.	C	79.	A	108.	C
22.	A	51.	A	80.	B	109.	B
23.	B	52.	A	81.	B	110.	B
24.	A	53.	B	82.	C	111.	A
25.	A	54.	B	83.	D	112.	A
26.	D	55.	C	84.	D	113.	D
27.	C	56.	A	85.	B	114.	D
28.	D	57.	C	86.	A	115.	B
29.	B	58.	D	87.	C	116.	C

©All rights reserved-Math-Knots LLC., VA-USA

www.math-knots.com | www.a4ace.com

Grade 8

Vol 1 Answer Key

Week 2
117.	D
118.	D
119.	A
120.	A

Week 3
1.	C
2.	C
3.	B
4.	A
5.	A
6.	C
7.	D
8.	B
9.	C
10.	B
11.	A
12.	D
13.	D
14.	A
15.	C
16.	B
17.	C
18.	C
19.	A
20.	C
21.	D
22.	B
23.	A
24.	C
25.	A
26.	A
27.	B
28.	A
29.	C

Week 3
30.	B
31.	A
32.	C
33.	C
34.	A
35.	C
36.	C
37.	B
38.	A
39.	A
40.	C
41.	A
42.	A
43.	B
44.	B
45.	D
46.	D
47.	C
48.	B
49.	C
50.	B
51.	C
52.	C
53.	C
54.	C
55.	D
56.	A
57.	A
58.	B

Week 3
59.	A
60.	B
61.	D
62.	B
63.	A
64.	A
65.	A
66.	C
67.	D
68.	C
69.	A
70.	D
71.	D
72.	A
73.	B
74.	D
75.	C
76.	D
77.	A
78.	A
79.	C
80.	C
81.	D
82.	A
83.	C
84.	D
85.	B
86.	D
87.	C

Grade 8

Vol 1 Answer Key

Week 3		Week 3		Week 4		Week 4	
88.	C	117.	D	1.	7 & 8	30.	29 & 30
89.	C	118.	D	2.	1 & 2	31.	25 & 26
90.	D	119.	C	3.	6 & 7	32.	16 & 17
91.	B	120.	A	4.	6 & 7	33.	29 & 30
92.	B	121.	D	5.	1 & 2	34.	10 & 11
93.	B	122.	A	6.	5 & 6	35.	28
94.	D	123.	B	7.	8 & 9	36.	28 & 29
95.	B	124.	A	8.	2 & 3	37.	14 & 15
96.	A	125.	B	9.	9 & 10	38.	22 & 23
97.	A			10.	7 & 8	39.	25 & 26
98.	D			11.	6 & 7	40.	18 & 19
99.	B			12.	7 & 8	41.	18
100.	A			13.	6 & 7	42.	19
101.	B			14.	3 & 4	43.	4
102.	A			15.	6 & 7	44.	-27
103.	B			16.	8 & 9	45.	-11
104.	D			17.	8 & 9	46.	-21
105.	A			18.	8 & 9	47.	-8
106.	A			19.	7	48.	24
107.	A			20.	3 & 4	49.	-14
108.	A			21.	7 & 8	50.	25
109.	C			22.	6 & 7	51.	8
110.	D			23.	9 & 6	52.	-13
111.	A			24.	7 & 8	53.	-1
112.	C			25.	7 & 8	54.	-29
113.	A			26.	26 & 27	55.	-2
114.	B			27.	28 & 29	56.	-14
115.	D			28.	10 & 11	57.	-7
116.	A			29.	18 & 19	58.	12

©All rights reserved-Math-Knots LLC., VA-USA

Grade 8

Vol 1 Answer Key

Week 4		Week 4		Week 4		Week 5	
59.	-23	77.	C	105.	D	1.	A
60.	-3	78.	C	106.	A	2.	D
61.	-20	79.	A	107.	C	3.	A
62.	-6	80.	A	108.	B	4.	A
63.	27	81.	A	109.	D	5.	C
64.	-7	82.	C	110.	D	6.	C
65.	-5	83.	A	111.	D	7.	D
66.	$-\frac{3}{8}$	84.	C	112.	B	8.	C
		85.	A	113.	C	9.	D
67.	$-\frac{2}{5}$	86.	D	114.	A	10.	A
		87.	C	115.	C	11.	D
68.	$-\frac{7}{8}$	88.	C	116.	D	12.	A
		89.	D	117.	A	13.	B
69.	$-\frac{9}{11}$	90.	B	118.	B	14.	D
		91.	A	119.	A	15.	A
70.	$\frac{3}{5}$	92.	B	120.	C	16.	C
		93.	C	121.	A	17.	B
71.	$-\frac{1}{7}$	94.	A	122.	C	18.	C
		95.	B	123.	C	19.	B
72.	$\frac{1}{3}$	96.	D	124.	B	20.	B
		97.	A			21.	A
73.	$-\frac{2}{3}$	98.	C			22.	A
		99.	B			23.	A
74.	$-\frac{5}{9}$	100.	C			24.	A
		101.	B			25.	B
75.	$-\frac{9}{10}$	102.	C			26.	B
		103.	D			27.	B
76.	$-\frac{1}{50}$	104.	D			28.	A
						29.	B

Grade 8

Vol 1 Answer Key

Week 5		Week 5		Week 5		Week 5	
30.	D	59.	C	88.	D	117.	A
31.	B	60.	B	89.	A	118.	C
32.	A	61.	C	90.	C	119.	A
33.	B	62.	A	91.	A	120.	B
34.	A	63.	D	92.	A	121.	B
35.	A	64.	C	93.	A	122.	C
36.	A	65.	A	94.	A	123.	C
37.	D	66.	B	95.	D	124.	A
38.	D	67.	D	96.	C	125.	C
39.	D	68.	C	97.	A	126.	C
40.	C	69.	D	98.	C	127.	D
41.	B	70.	A	99.	D	128.	A
42.	D	71.	B	100.	A	129.	A
43.	D	72.	A	101.	A	130.	A
44.	C	73.	B	102.	B	131.	D
45.	B	74.	D	103.	C	132.	B
46.	D	75.	C	104.	D	133.	B
47.	B	76.	A	105.	A	134.	B
48.	A	77.	D	106.	A	135.	D
49.	D	78.	C	107.	C		
50.	A	79.	A	108.	B		
51.	A	80.	D	109.	A		
52.	A	81.	B	110.	B		
53.	A	82.	D	111.	A		
54.	A	83.	A	112.	A		
55.	A	84.	C	113.	A		
56.	D	85.	B	114.	A		
57.	B	86.	B	115.	A		
58.	C	87.	D	116.	D		

©All rights reserved-Math-Knots LLC., VA-USA
www.math-knots.com | www.a4ace.com

Grade 8

Vol 1 Answer Key

Week 6		Week 6		Week 6		Week 6	
1.	D	30.	D	59.	B	88.	D
2.	B	31.	A	60.	A	89.	B
3.	C	32.	D	61.	B	90.	C
4.	D	33.	D	62.	B	91.	D
5.	C	34.	A	63.	B	92.	C
6.	C	35.	D	64.	D	93.	B
7.	C	36.	D	65.	C	94.	D
8.	B	37.	A	66.	A	95.	B
9.	B	38.	B	67.	A	96.	A
10.	B	39.	B	68.	D	97.	B
11.	B	40.	D	69.	A	98.	A
12.	A	41.	C	70.	C	99.	A
13.	A	42.	B	71.	A	100.	D
14.	A	43.	D	72.	C	101.	D
15.	C	44.	C	73.	A	102.	D
16.	A	45.	A	74.	C	103.	A
17.	D	46.	A	75.	D	104.	C
18.	C	47.	C	76.	C	105.	A
19.	A	48.	A	77.	C	106.	A
20.	D	49.	C	78.	A	107.	A
21.	B	50.	A	79.	B	108.	D
22.	C	51.	B	80.	A	109.	D
23.	D	52.	C	81.	D	110.	D
24.	A	53.	C	82.	B	111.	A
25.	A	54.	A	83.	B	112.	C
26.	B	55.	C	84.	A	113.	C
27.	D	56.	D	85.	A	114.	B
28.	C	57.	A	86.	D	115.	B
29.	C	58.	D	87.	D	116.	C

Grade 8

Vol 1 Answer Key

Week 6		Week 7		Week 7		Week 7	
117.	A	1.	D	30.	A	59.	A
118.	B	2.	C	31.	D	60.	A
119.	C	3.	B	32.	C	61.	B
120.	A	4.	D	33.	C	62.	A
121.	D	5.	D	34.	A	63.	B
122.	C	6.	D	35.	D	64.	A
123.	A	7.	C	36.	C	65.	A
124.	A	8.	D	37.	B	66.	B
125.	C	9.	D	38.	A	67.	D
126.	B	10.	D	39.	D	68.	B
127.	B	11.	B	40.	A	69.	A
128.	C	12.	A	41.	B	70.	A
129.	C	13.	B	42.	D	71.	C
130.	B	14.	C	43.	B	72.	A
131.	B	15.	C	44.	C	73.	D
132.	D	16.	A	45.	C	74.	D
133.	D	17.	D	46.	C	75.	C
134.	C	18.	D	47.	A	76.	C
135.	A	19.	C	48.	C	77.	C
136.	D	20.	D	49.	B	78.	A
137.	D	21.	A	50.	C	79.	C
138.	B	22.	C	51.	A	80.	B
139.	A	23.	C	52.	B	81.	A
140.	B	24.	A	53.	C	82.	D
		25.	B	54.	D	83.	C
		26.	A	55.	A	84.	A
		27.	A	56.	B	85.	C
		28.	D	57.	A	86.	D
		29.	D	58.	C	87.	D

©All rights reserved-Math-Knots LLC., VA-USA

www.math-knots.com | www.a4ace.com

Grade 8

Vol 1 Answer Key

Week 7

88.	B
89.	D
90.	A
91.	C

Week 8 Assessment 1

1.	25 & 26
2.	26 & 27
3.	B
4.	B
5.	D
6.	A
7.	-5
8.	B
9.	C
10.	B
11.	A
12.	C
13.	D
14.	B
15.	D
16.	D
17.	$\frac{-1}{30}$
18.	A
19.	D
20.	C
21.	A
22.	C
23.	$\frac{1}{16}$
24.	A
25.	D
26.	B
27.	C

Week 8 Assessment 1

28.	D
29.	B
30.	$\frac{-1}{4}$
31.	A
32.	C
33.	D
34.	D
35.	B
36.	B
37.	A
38.	C
39.	B
40.	A
41.	B
42.	B
43.	B
44.	A
45.	A
46.	D
47.	D
48.	A
49.	D
50.	D
51.	C
52.	A
53.	D
54.	D
55.	D

Week 8 Assessment 1

56.	D
57.	C
58.	B
59.	A
60.	A
61.	B
62.	C
63.	D
64.	D
65.	D
66.	A
67.	C
68.	B
69.	C
70.	D
71.	C
72.	C

©All rights reserved-Math-Knots LLC., VA-USA   www.math-knots.com | www.a4ace.com

Grade 8

Vol 1 Answer Key

Week 9		Week 9		Week 9		Week 9	
1.	B	30.	C	59.	A	88.	A
2.	C	31.	C	60.	B	89.	A
3.	B	32.	C	61.	D	90.	B
4.	D	33.	D	62.	D	91.	A
5.	B	34.	C	63.	B	92.	A
6.	A	35.	D	64.	A	93.	B
7.	D	36.	B	65.	D	94.	D
8.	A	37.	C	66.	B	95.	B
9.	A	38.	A	67.	A	96.	A
10.	A	39.	B	68.	D	97.	C
11.	A	40.	D	69.	B	98.	C
12.	B	41.	B	70.	A	99.	C
13.	A	42.	C	71.	B	100.	A
14.	D	43.	D	72.	D	101.	D
15.	B	44.	B	73.	A	102.	C
16.	A	45.	B	74.	D	103.	B
17.	B	46.	A	75.	C	104.	A
18.	C	47.	B	76.	A	105.	A
19.	A	48.	C	77.	A	106.	D
20.	C	49.	A	78.	A	107.	B
21.	C	50.	B	79.	B	108.	C
22.	A	51.	C	80.	D	109.	B
23.	D	52.	C	81.	A	110.	C
24.	A	53.	B	82.	B	111.	D
25.	A	54.	D	83.	C	112.	D
26.	C	55.	D	84.	D	113.	A
27.	B	56.	A	85.	D	114.	A
28.	D	57.	D	86.	C	115.	D
29.	B	58.	B	87.	A	116.	D

©All rights reserved-Math-Knots LLC., VA-USA

www.math-knots.com | www.a4ace.com

Grade 8

Vol 1 Answer Key

Week 9		Week 9		Week 10		Week 10	
117.	B	146.	B	1.	A	30.	A
118.	B	147.	D	2.	B	31.	B
119.	A	148.	A	3.	B	32.	C
120.	D	149.	A	4.	C	33.	C
121.	A	150.	C	5.	D	34.	C
122.	B	151.	D	6.	C	35.	C
123.	C	152.	A	7.	B	36.	C
124.	A	153.	A	8.	D	37.	C
125.	D	154.	A	9.	C	38.	D
126.	D	155.	D	10.	B	39.	A
127.	A	156.	B	11.	B	40.	A
128.	D	157.	D	12.	D	41.	A
129.	B	158.	C	13.	A	42.	B
130.	D	159.	B	14.	C	43.	B
131.	C	160.	B	15.	C	44.	C
132.	B			16.	B	45.	A
133.	D			17.	C	46.	B
134.	B			18.	D	47.	B
135.	D			19.	A	48.	B
136.	A			20.	C	49.	B
137.	A			21.	B	50.	A
138.	A			22.	A	51.	D
139.	D			23.	C	52.	D
140.	A			24.	A	53.	A
141.	A			25.	D	54.	A
142.	D			26.	C	55.	B
143.	B			27.	B	56.	D
144.	A			28.	D	57.	C
145.	B			29.	B	58.	D

Grade 8

Vol 1 Answer Key

Week 10		Week 10		Week 10		Week 11	
59.	C	88.	C	117.	A	1.	B
60.	C	89.	C	118.	D	2.	A
61.	C	90.	C	119.	B	3.	B
62.	D	91.	C	120.	C	4.	D
63.	C	92.	D	121.	C	5.	A
64.	A	93.	D	122.	C	6.	B
65.	A	94.	B	123.	A	7.	A
66.	D	95.	D	124.	A	8.	A
67.	A	96.	A	125.	D	9.	C
68.	A	97.	B	126.	B	10.	A
69.	D	98.	D	127.	C	11.	C
70.	A	99.	A	128.	C	12.	C
71.	B	100.	D	129.	A	13.	B
72.	A	101.	A	130.	D	14.	C
73.	D	102.	D	131.	A	15.	C
74.	C	103.	B	132.	A	16.	C
75.	C	104.	A	133.	A	17.	A
76.	A	105.	B	134.	C	18.	B
77.	A	106.	C	135.	C	19.	C
78.	C	107.	B	136.	A	20.	A
79.	D	108.	B	137.	C	21.	C
80.	B	109.	C	138.	B	22.	D
81.	A	110.	B			23.	B
82.	B	111.	C			24.	C
83.	B	112.	D			25.	A
84.	B	113.	C			26.	B
85.	B	114.	A			27.	D
86.	A	115.	D			28.	B
87.	D	116.	D			29.	D

©All rights reserved-Math-Knots LLC., VA-USA   www.math-knots.com | www.a4ace.com

Grade 8

Vol 1 Answer Key

Week 11

30.	C
31.	B
32.	C
33.	C
34.	A
35.	D
36.	B
37.	C
38.	C
39.	C
40.	D
41.	C
42.	B
43.	B
44.	A
45.	D
46.	C
47.	B
48.	C
49.	C
50.	A
51.	A
52.	D
53.	B
54.	C
55.	C
56.	A
57.	C
58.	D

Week 11

59.	B
60.	A
61.	D
62.	A
63.	B
64.	B
65.	D
66.	6 & 9
67.	21 & 28
68.	16
69.	A
70.	6, 12 & 18
71.	28 & 30
72.	16 & 12
73.	18 years & 46 years
74.	196
75.	12, 30 & 48
76.	104 & 39
77.	4, 6 & 8
78.	6 & 9
79.	76, 78 & 80
80.	40, 50 & 60
81.	96
82.	$\frac{5}{8}$
83.	44, 55 & 66
84.	10 years & 40 years

Week 11

85.	60
86.	57 mph & 62 mph
87.	17 & 51
88.	70 mph & 63 mph
89.	D
90.	A
91.	B
92.	A
93.	D
94.	D
95.	C
96.	A
97.	D
98.	A
99.	D
100.	D
101.	B
102.	C
103.	A
104.	A
105.	D
106.	D
107.	A
108.	A
109.	D
110.	A
111.	B

Week 11

112.	D
113.	A
114.	B
115.	D
116.	D
117.	D
118.	D
119.	A
120.	B
121.	B
122.	A

Grade 8

Vol 1 Answer Key

Week 12

#	Ans
1.	B
2.	C
3.	C
4.	C
5.	D
6.	D
7.	A
8.	D
9.	D
10.	C
11.	B
12.	A
13.	A
14.	D
15.	D
16.	B
17.	A
18.	C
19.	D
20.	C
21.	B
22.	B
23.	D
24.	D
25.	C
26.	C
27.	B
28.	A
29.	A

Week 12

#	Ans
30.	A
31.	A
32.	A
33.	C
34.	A
35.	B
36.	D
37.	C
38.	D
39.	D
40.	D
41.	B
42.	C
43.	D
44.	C
45.	A
46.	C
47.	C
48.	B
49.	D
50.	C
51.	C
52.	C
53.	B
54.	A
55.	D
56.	D
57.	D
58.	A

Week 12

#	Ans
59.	C
60.	A
61.	B
62.	B
63.	C
64.	A
65.	(-4, 2, 5)
66.	(-2, 3.5)
67.	(0, 2)
68.	(1, -5)
69.	(-1.5, -2.5)
70.	(-0.5, 1)
71.	(0, -2.5)
72.	(-2, -3)
73.	(1, -2.5)
74.	(2, 8)
75.	(4, 6)
76.	(11.5, 2)
77.	(4.5, -1.5)
78.	(0, 11.5)
79.	(-3, -3)
80.	(-9, 4)
81.	(10.5, -1)
82.	(6, 7)
83.	(-11, -7)
84.	(7, -1.5)
85.	(-9, -3)
86.	(6, 4.5)
87.	(-9, 8)

Week 12

#	Ans
88.	(11, 9)
89.	(17, -8)
90.	(-4, -3)
91.	(6.5, 6.5)
92.	(7.5, 10)
93.	(-1, -13)
94.	(15, -9)
95.	(18.5, -3)
96.	(13.5, -6)
97.	(3.5, 1)
98.	(-14, -7.5)
99.	(3.5, -11)
100.	(-13.5, 3.5)
101.	C
102.	C
103.	B
104.	A
105.	D
106.	C
107.	B
108.	D

Grade 8

Vol 1 Answer Key

Week 13

#	Ans
1.	A
2.	C
3.	B
4.	C
5.	C
6.	A
7.	A
8.	C
9.	D
10.	B
11.	B
12.	C
13.	C
14.	B
15.	C
16.	C
17.	D
18.	A
19.	B
20.	C
21.	A
22.	A
23.	D
24.	A
25.	B
26.	A
27.	B
28.	D
29.	A

Week 13

#	Ans
30.	D
31.	A
32.	A
33.	C
34.	B
35.	B
36.	B
37.	B
38.	C
39.	B
40.	B
41.	D
42.	A
43.	D
44.	B
45.	B
46.	B
47.	B
48.	C
49.	C
50.	D
51.	C
52.	B
53.	D
54.	A
55.	C
56.	B
57.	B
58.	D

Week 13

#	Ans
59.	C
60.	B
61.	C
62.	B
63.	D
64.	C
65.	D
66.	B
67.	D
68.	B
69.	C
70.	C
71.	C
72.	C
73.	C
74.	B
75.	D
76.	B
77.	A
78.	B
79.	D
80.	C
81.	B
82.	B
83.	C
84.	A
85.	D
86.	C
87.	B

Week 13

#	Ans
88.	A
89.	A
90.	C
91.	D
92.	C
93.	71
94.	87
95.	47
96.	51
97.	70
98.	19
99.	41
100.	28
101.	50
102.	36
103.	76
104.	65
105.	B
106.	A
107.	A
108.	C
109.	C
110.	D
111.	D
112.	D
113.	C
114.	C
115.	D
116.	A

Grade 8

Vol 1 Answer Key

Week 13

117.	A
118.	D
119.	C
120.	A
121.	D
122.	A
123.	A
124.	C
125.	A
126.	B
127.	D

Week 14

1.	A
2.	A
3.	A
4.	B
5.	B
6.	D
7.	B
8.	C
9.	A
10.	C
11.	C
12.	D
13.	D
14.	C
15.	A
16.	C
17.	D
18.	B
19.	B
20.	D
21.	D
22.	D
23.	D
24.	C
25.	D
26.	D
27.	D
28.	A
29.	D

Week 14

30.	B
31.	A
32.	D
33.	B
34.	A
35.	A
36.	D
37.	D
38.	A
39.	C
40.	C
41.	B
42.	D
43.	B
44.	A
45.	D
46.	D
47.	B
48.	D
49.	D
50.	B
51.	B
52.	D
53.	C
54.	D
55.	C
56.	D
57.	D
58.	C

Week 14

59.	A
60.	C
61.	D
62.	A

Grade 8

Vol 1 Answer Key

63. B

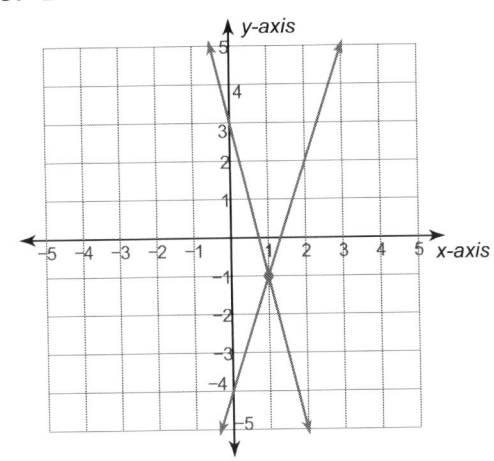

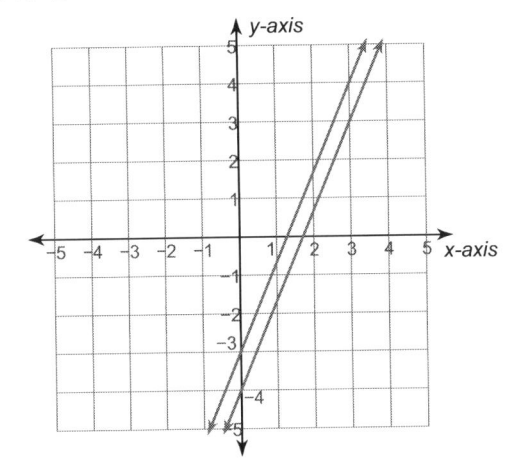

64. D

65. A

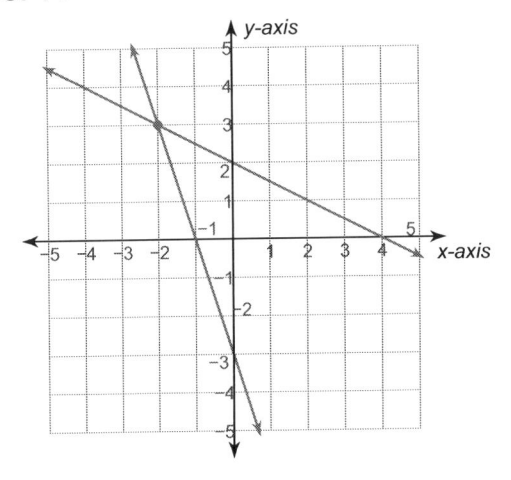

66. B

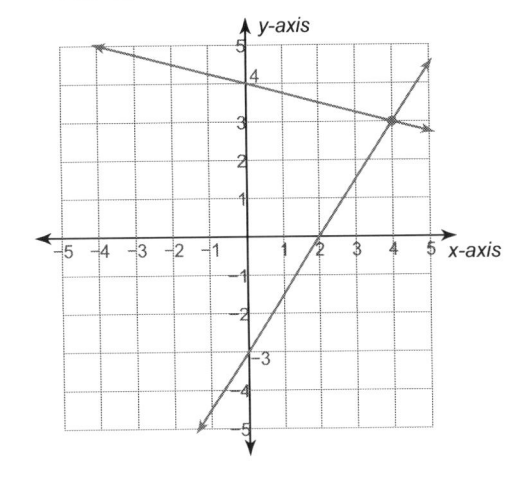

67. A

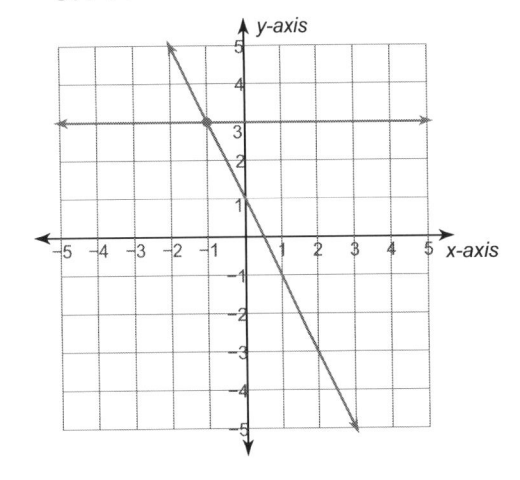

68. B

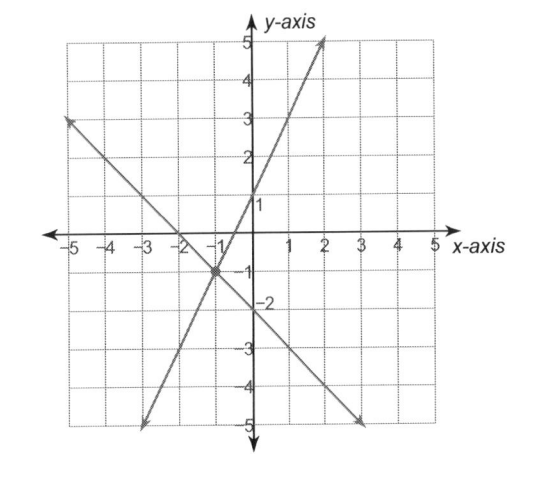

295

Grade 8

Vol 1 Answer Key

69. A

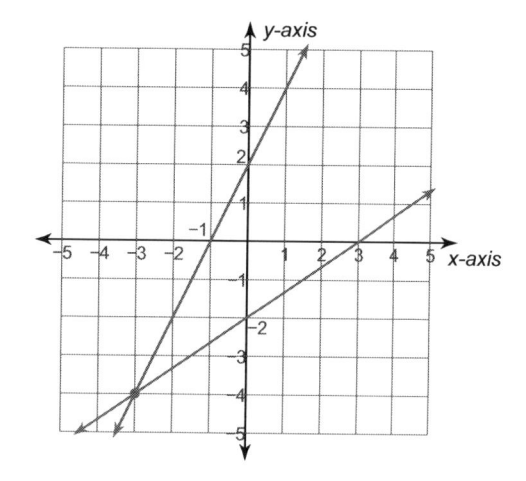

70. B

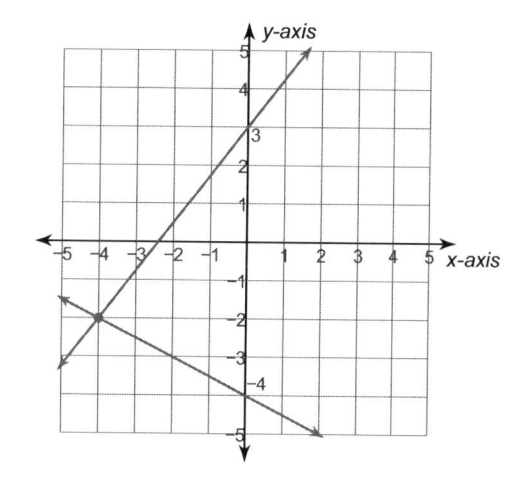

71. B

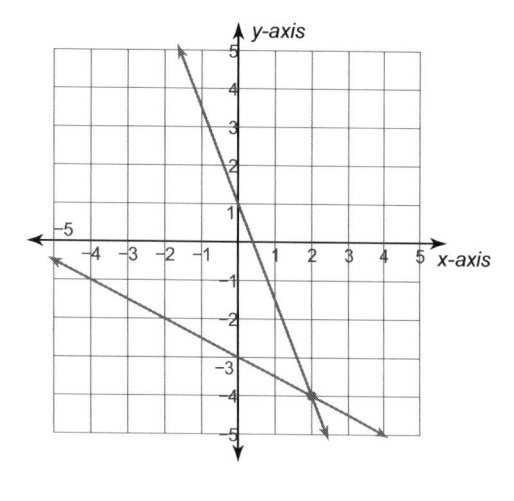

72. A

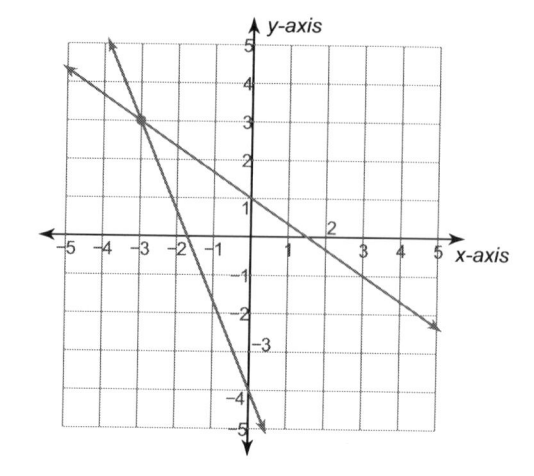

73. A

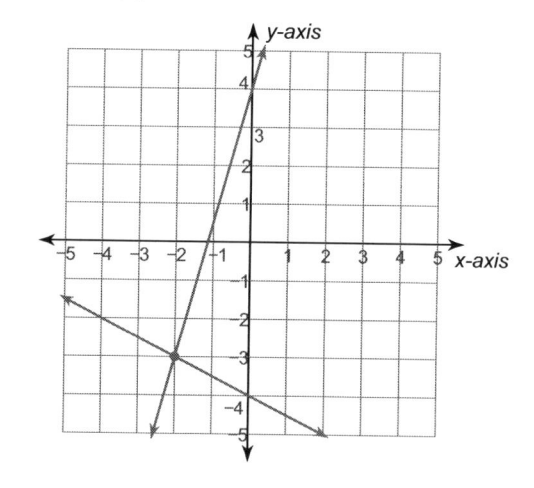

74. C

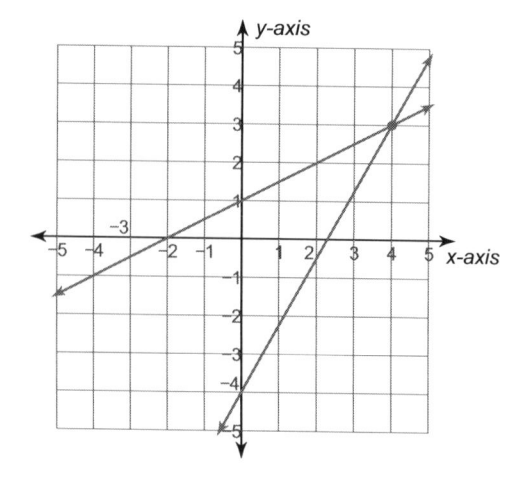

©All rights reserved-Math-Knots LLC., VA-USA   296   www.math-knots.com | www.a4ace.com

Grade 8

Vol 1 Answer Key

75. B

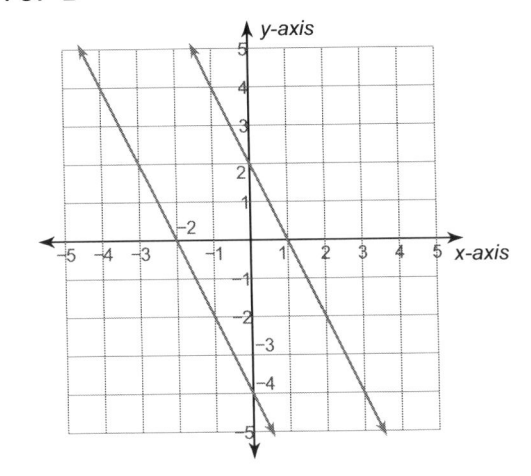

76. A

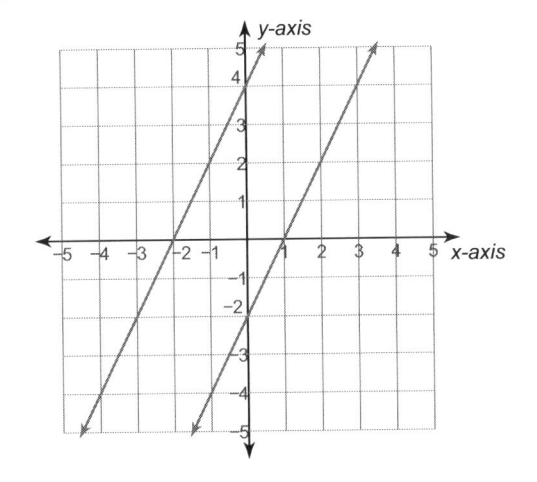

77. C

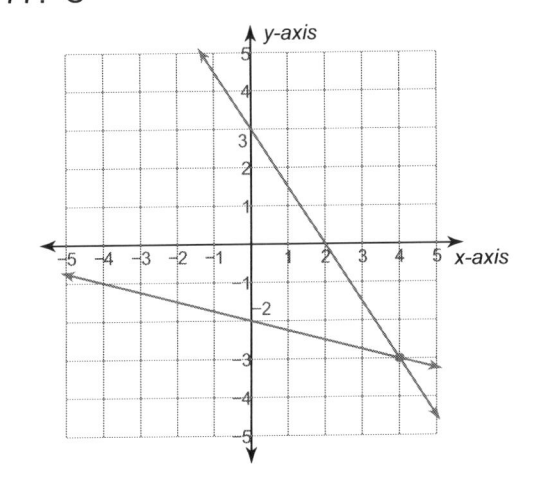

78. C

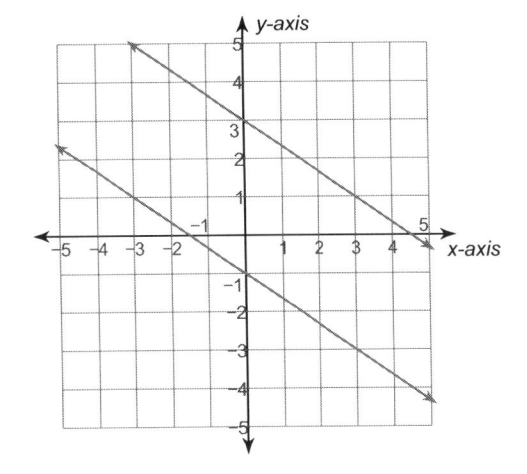

79. D

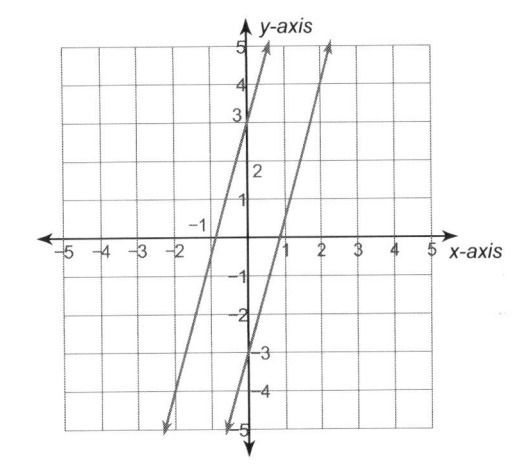

80. D

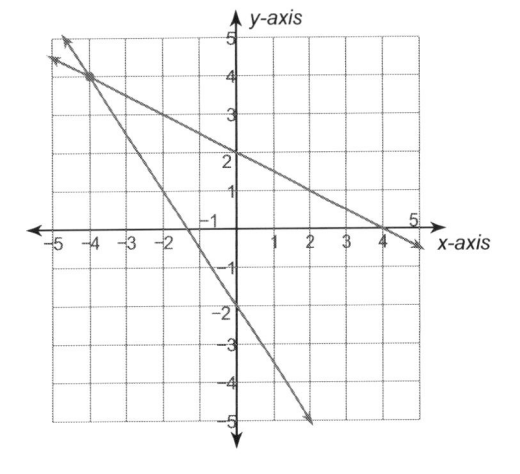

©All rights reserved-Math-Knots LLC., VA-USA 297 www.math-knots.com | www.a4ace.com

Grade 8

Vol 1 Answer Key

81. C

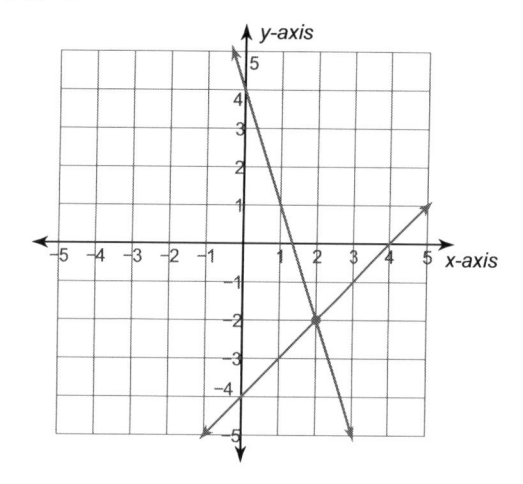

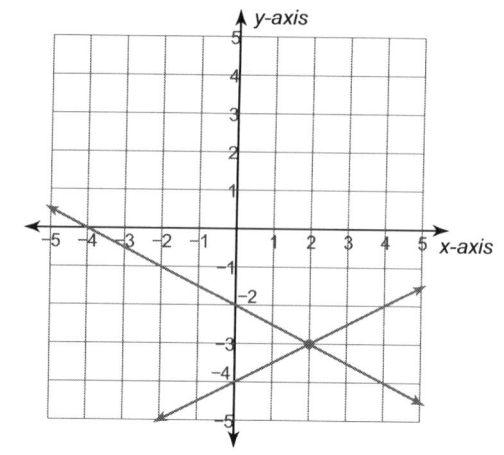

82. C

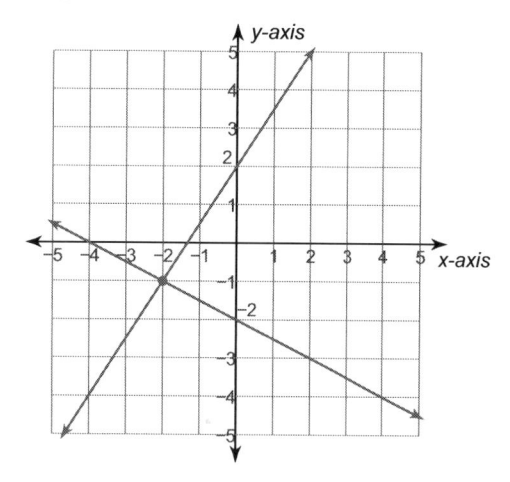

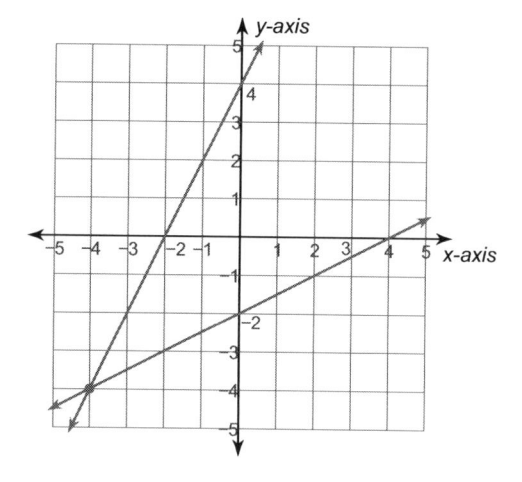

83. B

84. C

85. B

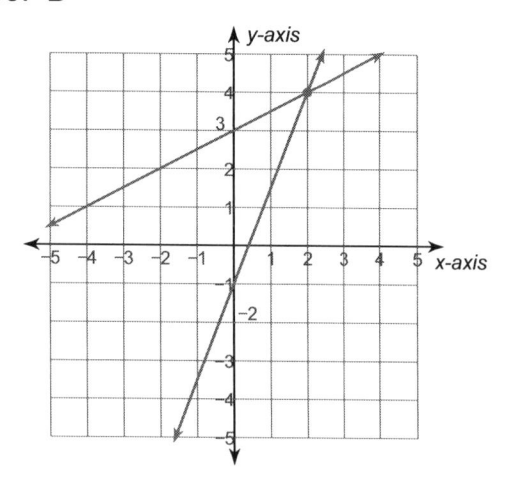

86. D

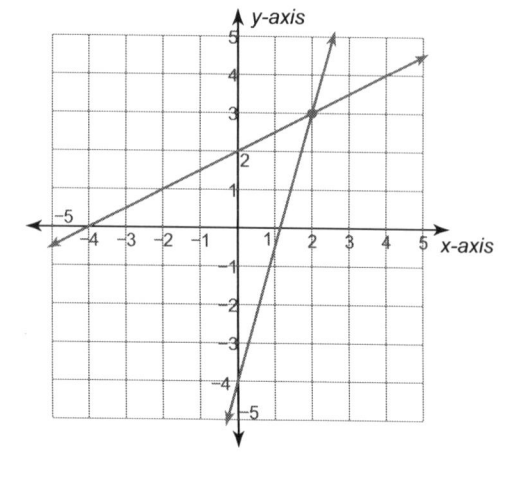

©All rights reserved-Math-Knots LLC., VA-USA

Grade 8

Vol 1 Answer Key

87. A

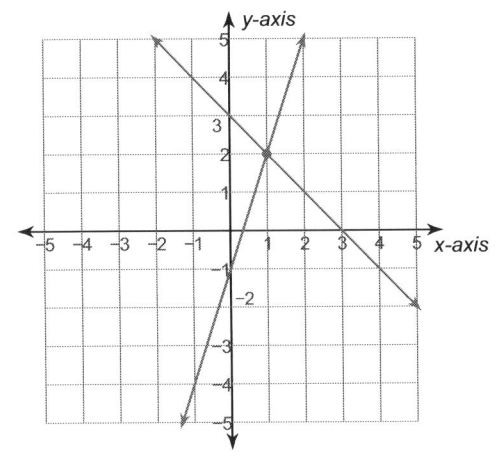

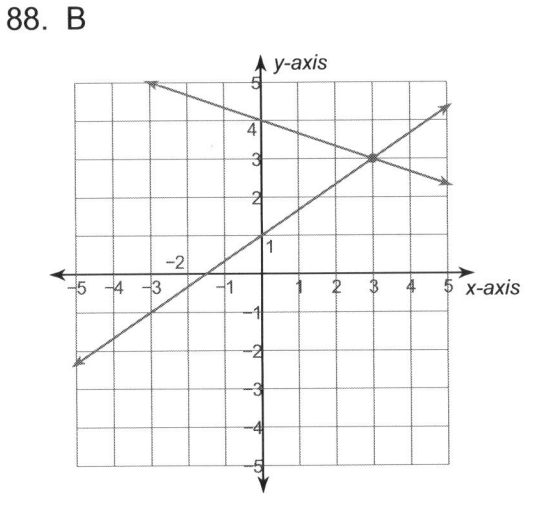

88. B

89. C

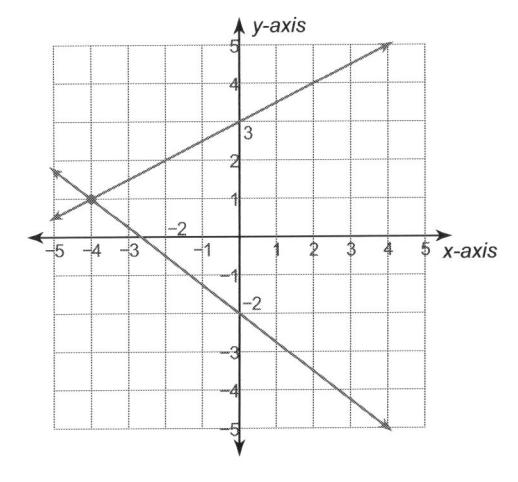

90. A

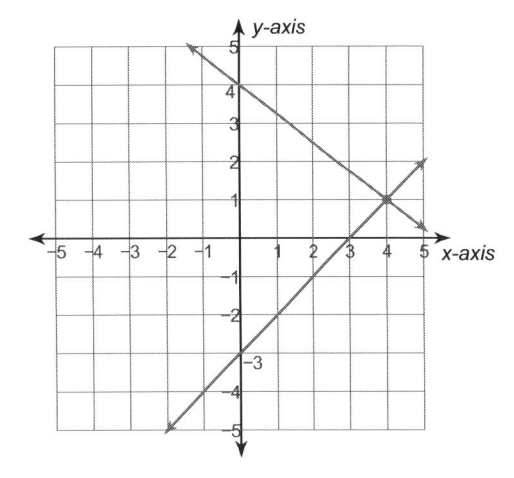

91. C

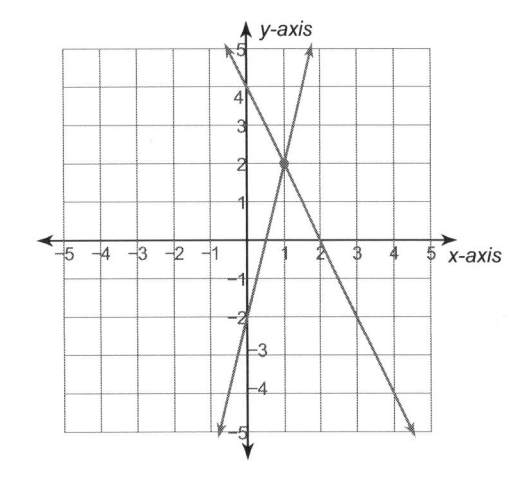

92. B

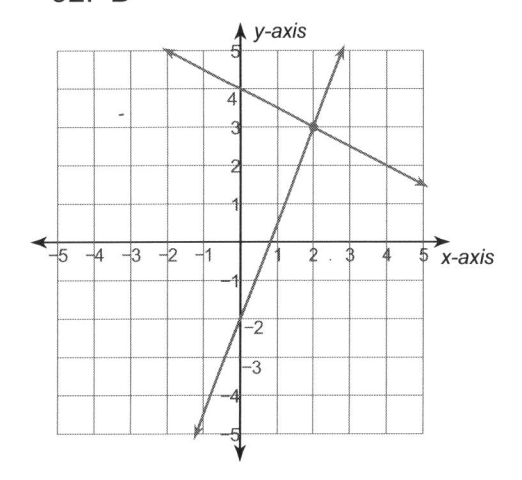

©All rights reserved-Math-Knots LLC., VA-USA   299   www.math-knots.com | www.a4ace.com

Grade 8

Vol 1 Answer Key

93. D

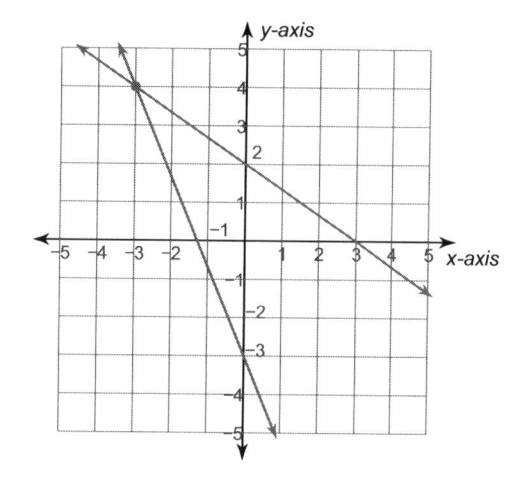

94. D

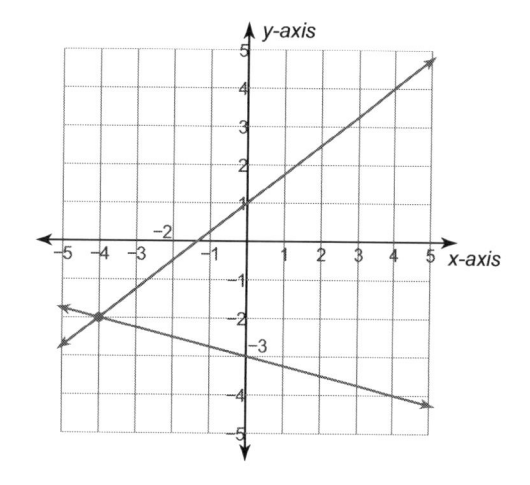

95. C

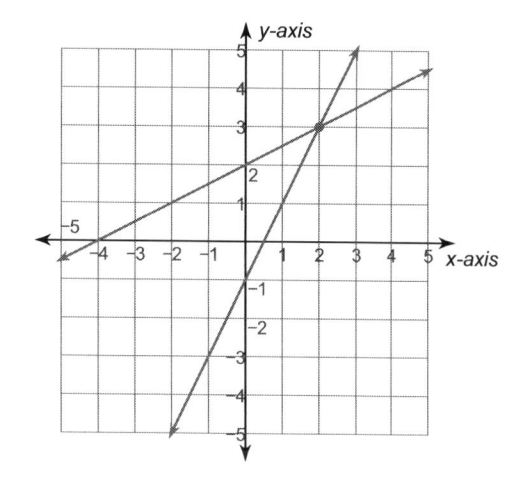

96. D

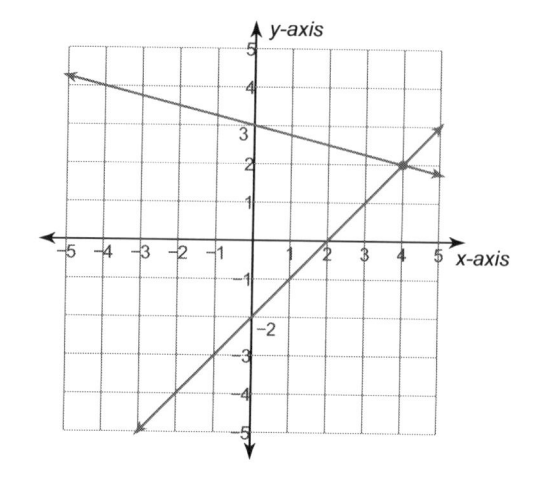

97. C

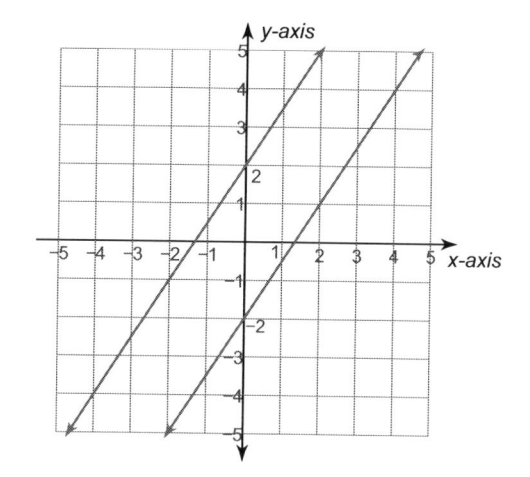

98. B

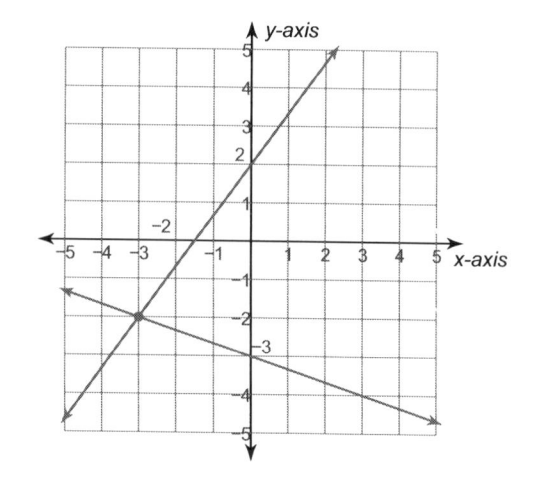

©All rights reserved-Math-Knots LLC., VA-USA

Vol 1 Answer Key

Grade 8

99. B

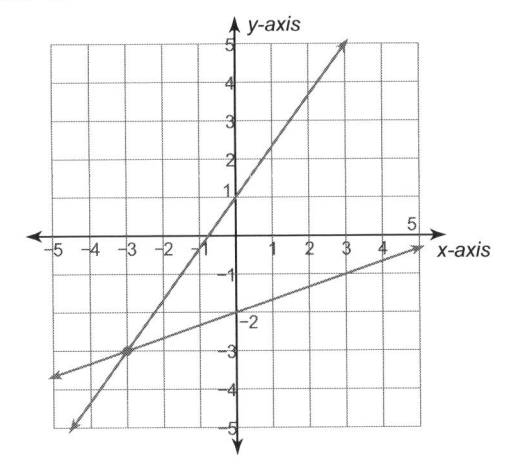

100. A

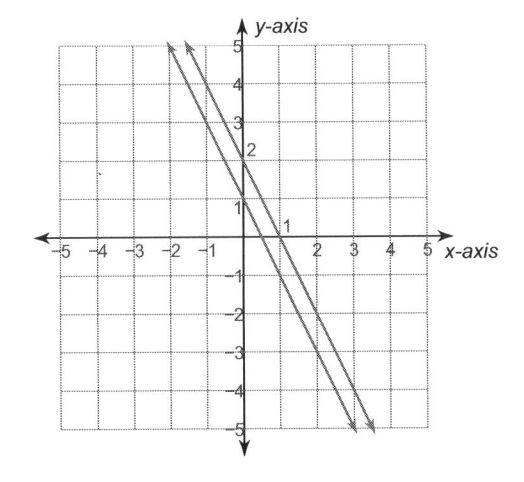

101. D

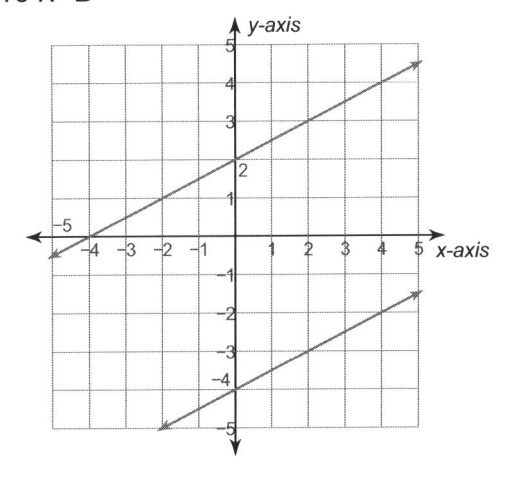

102. A

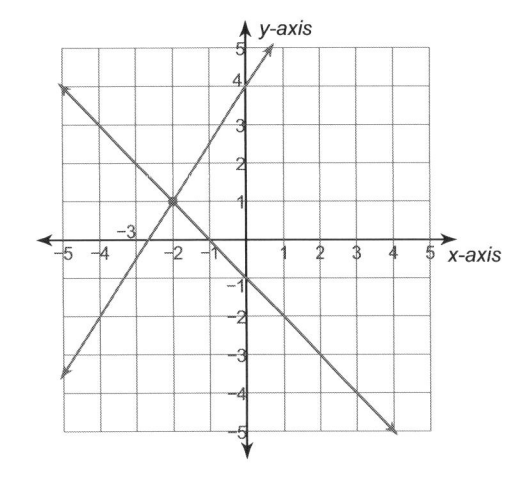

103. A

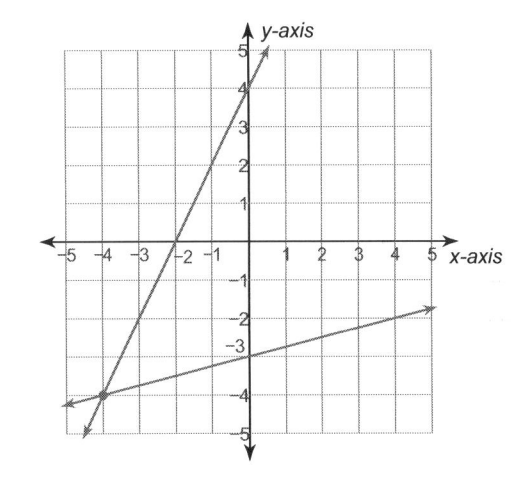

104. A

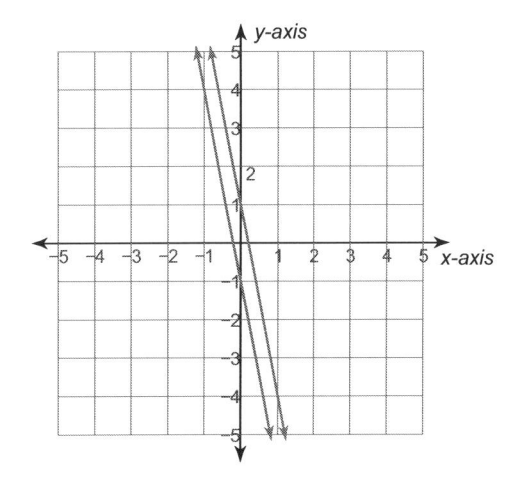

©All rights reserved-Math-Knots LLC., VA-USA   301   www.math-knots.com | www.a4ace.com

Grade 8

Vol 1 Answer Key

105. A

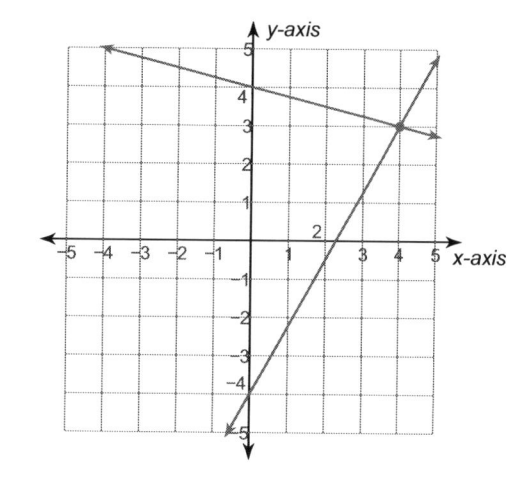

106. D

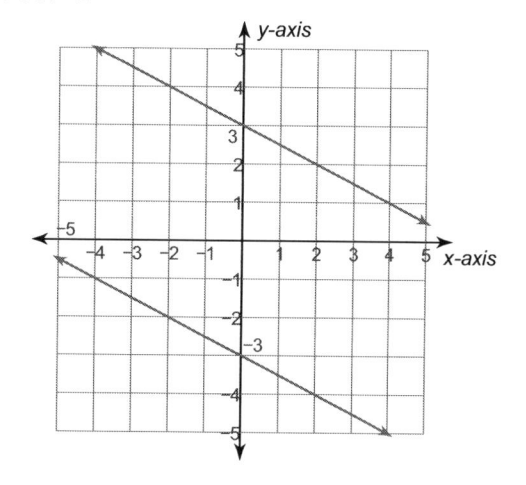

107. D

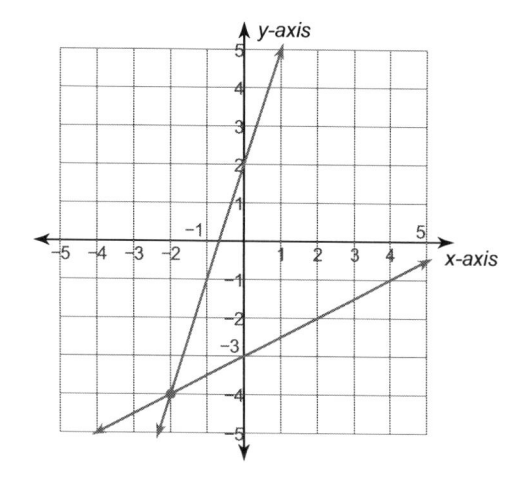

108. A

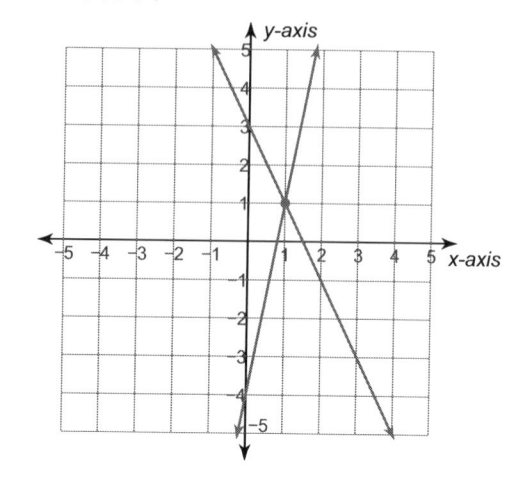

109. B

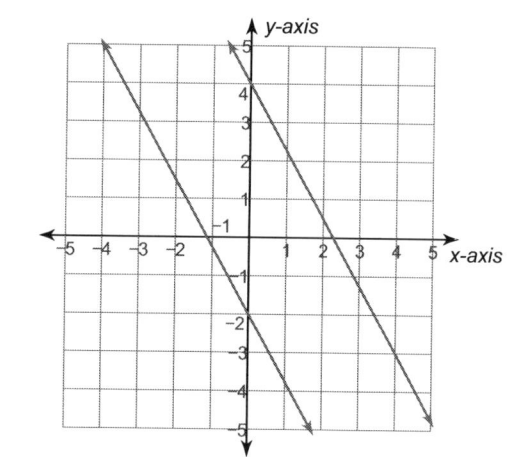

110. D

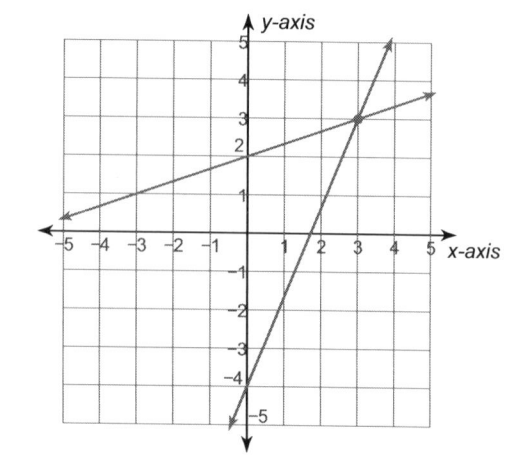

©All rights reserved-Math-Knots LLC., VA-USA   302   www.math-knots.com | www.a4ace.com

Grade 8

Vol 1 Answer Key

111. D

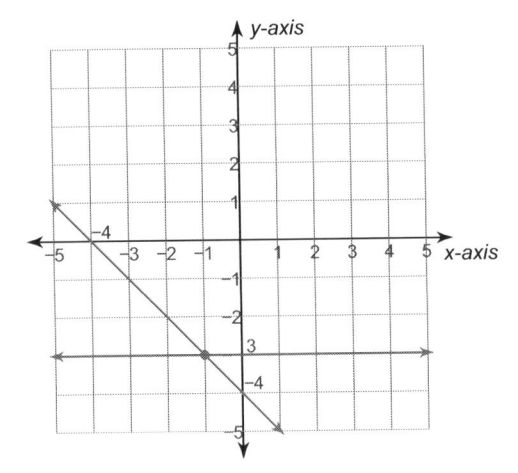

112. C

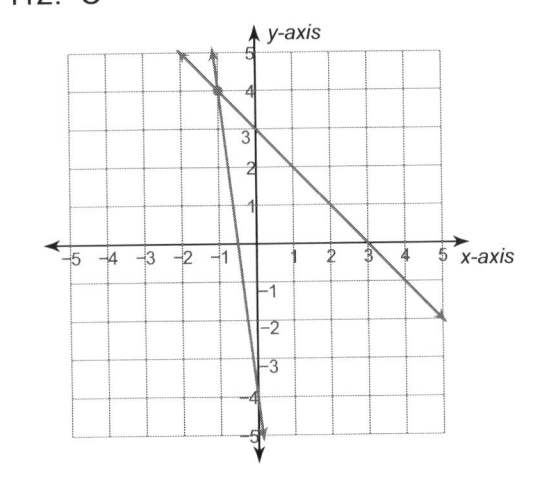

113. D

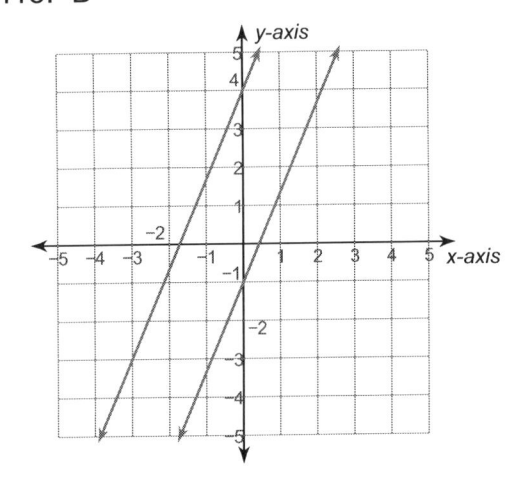

114. B

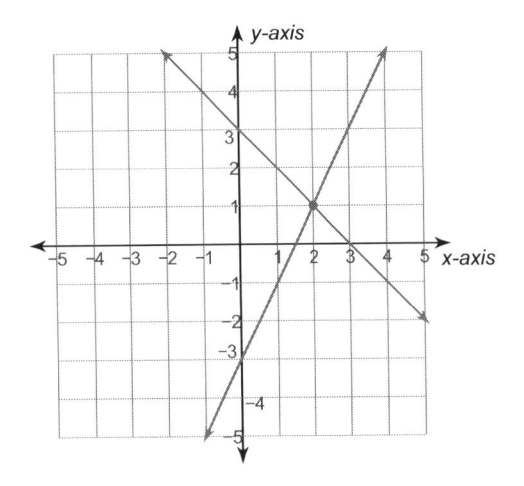

115. C

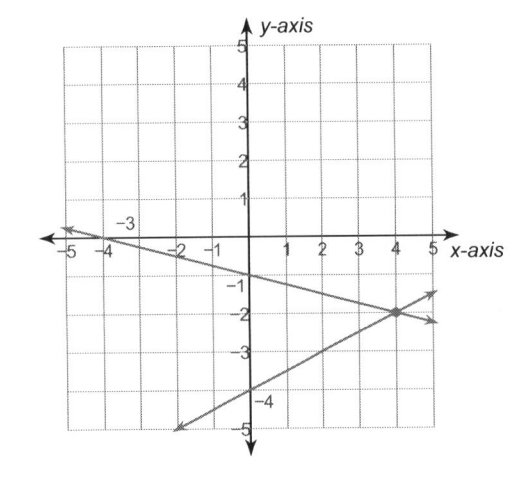

116. C

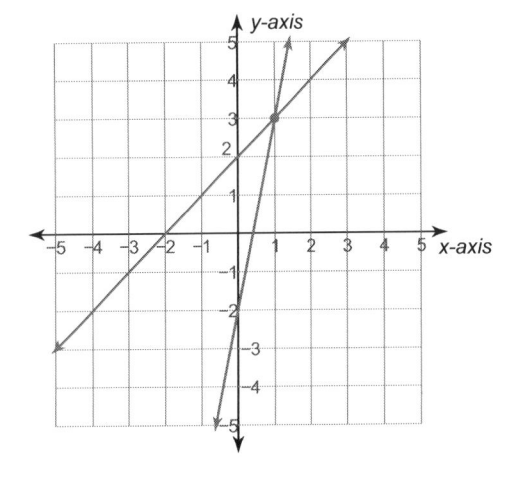

Grade 8

Vol 1 Answer Key

117. D

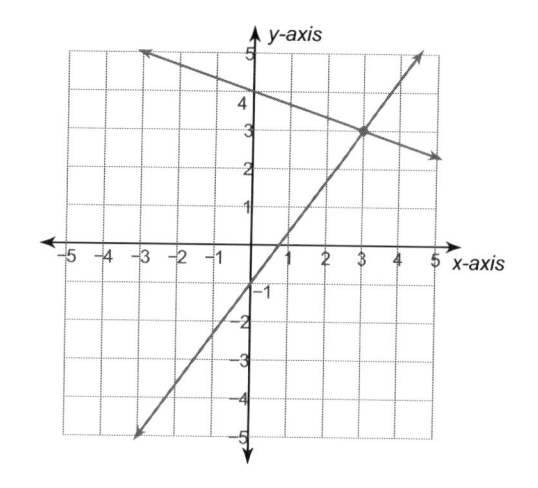

118. C

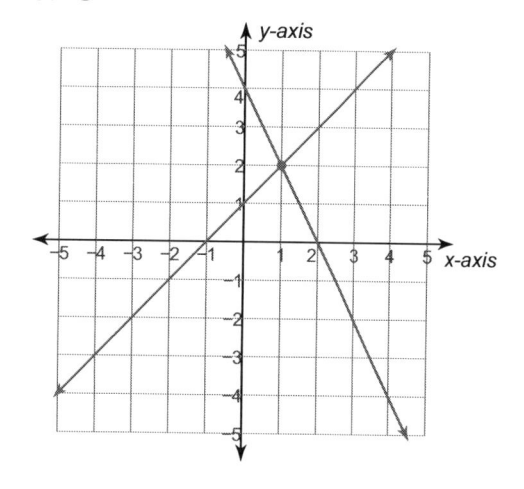

119. C

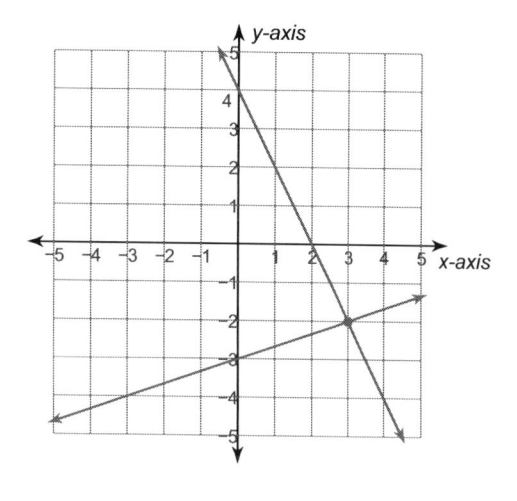

120. C

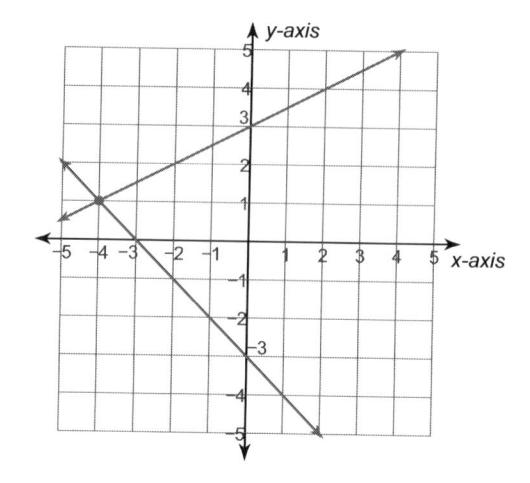

121. A

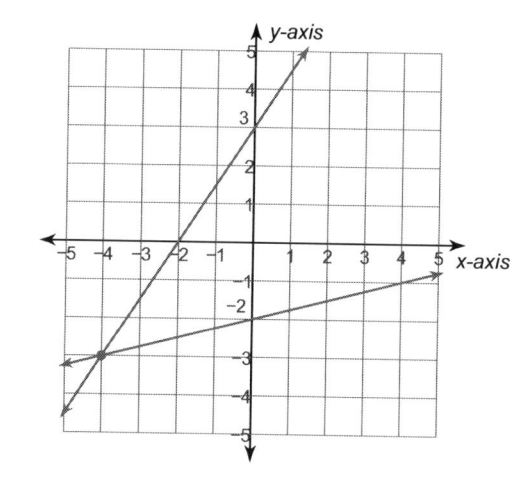

122. C

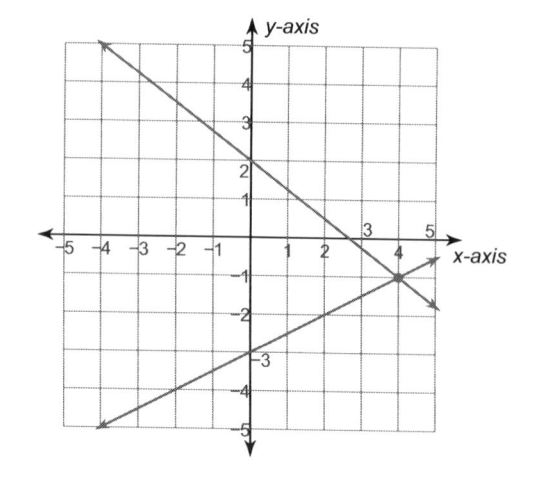

©All rights reserved-Math-Knots LLC., VA-USA   304   www.math-knots.com | www.a4ace.com

Grade 8

Vol 1 Answer Key

Week 14

123.	B
124.	A
125.	A
126.	C
127.	A
128.	B
129.	A
130.	A
131.	D
132.	D
133.	D
134.	A
135.	C
136.	D
137.	C
138.	D
139.	B
140.	D
141.	A
142.	C
143.	B
144.	B

Week 15

1.	B
2.	B
3.	A
4.	D
5.	C
6.	D
7.	B
8.	A
9.	A
10.	A
11.	B
12.	C
13.	A
14.	A
15.	C
16.	A
17.	A
18.	C
19.	A
20.	C
21.	A
22.	B
23.	B
24.	A
25.	C
26.	A
27.	(-14.5 , -5)
28.	B
29.	D

Week 15

30.	D
31.	B
32.	A
33.	B
34.	C
35.	A
36.	D
37.	C
38.	B
39.	D
40.	D
41.	C
42.	A
43.	C
44.	C
45.	A
46.	C
47.	A
48.	D
49.	C
50.	B
51.	A
52.	B
53.	(-0.5 , 4)
54.	(-3.5 , -8)
55.	C
56.	C
57.	A
58.	A

Grade 8

Vol 1 Answer Key

Week 15   **Week 15**

59.	D
60.	A
61.	A
62.	D
63.	B
64.	D
65.	C
66.	D
67.	A
68.	D
69.	C
70.	D
71.	C
72.	D
73.	B
74.	C
75.	C
76.	A
77.	D
78.	A
79.	D
80.	D
81.	B
82.	B
83.	B
84.	D
85.	C
86.	D
87.	C

88.	D
89.	A
90.	C
91.	B
92.	A
93.	B
94.	B
95.	D
96.	B
97.	A
98.	A
99.	71
100.	24
101.	97
102.	A
103.	A
104.	C
105.	D
106.	D
107.	$8
108.	$7

©All rights reserved-Math-Knots LLC., VA-USA   www.math-knots.com | www.a4ace.com

Made in the USA
Columbia, SC
05 December 2024